Whole Genome Amplification

The **METHODS EXPRESS** series

Series editor: B. David Hames

Faculty of Biological Sciences, University of Leeds, Leeds LS2 9JT, UK

Bioinformatics

Biosensors

Cell Imaging

DNA Microarrays

Immunohistochemistry

PCR

Protein Microarrays

Proteomics

METHODS EXPRESS

MX

Whole Genome Amplification

edited by S. Hughes

Institute of Cancer Research, London, UK

and R. Lasken

Center for Genomic Sciences, Allegheny-Singer Research Institute, Pittsburgh, USA

Scion

© Scion Publishing Ltd, 2005

First published 2005

A CIP catalogue record for this book is available from the British Library.

ISBN 1 904842 07 0 (paperback)
ISBN 1 904842 18 6 (hardback)

10064906399

Scion Publishing Limited
Bloxham Mill, Barford Road, Bloxham, Oxfordshire OX15 4FF
www.scionpublishing.com

Important Note from the Publisher

Typeset by Phoenix Photosetting, Chatham, Kent, UK
Printed by Biddles Ltd, King's Lynn, UK, www.biddles.co.uk

Cover image:
Microscopic isolation of single fluorescently labelled *Escherichia coli* cells by a micromanipulated microcapillary. The genome of the isolated cell is subsequently amplified by multiple-displacement amplification, as described in Chapter 9. Photograph taken by Thomas Kvist.

Contents

Contributors   ix
Preface   xiii
Abbreviations   xv

Color section   xvii

Chapter 1
Introduction to whole genome amplification
José M. Lage and Paul M. Lizardi
1. Introduction   1
2. Methods and approaches   2
3. References   8

Chapter 2
Single nucleotide polymorphism typing using degenerate-oligonucleotide-primed PCR-amplified products
Makoto Bannai and Katsushi Tokunaga
1. Introduction   11
2. Methods and approaches   12
    2.1. Methodology of DOP-PCR   12
    2.2. Applications of DOP-PCR   12
3. Recommended protocols   12
    3.1. Results   18
    3.2. Conclusion   21
4. References   21

Chapter 3
Whole genome amplification by improved primer-extension pre-amplification PCR
Peter J. Wild and Wolfgang Dietmaier
1. Introduction   23
    1.1. Method of choice   23
2. Methods and approaches   26
    2.1. Cell isolation   26

3. Recommended protocols   27
   3.1. Downstream applications   29
4. Troubleshooting   32
5. References   32

Chapter 4
Global amplification using SCOMP: single-cell comparative genomic hybridization
Nona C.R. Arneson, Arezou A. Ghazani and Susan J. Done
1. Introduction   33
2. Methods and approaches   34
   2.1. SCOMP   34
   2.2. Principles of SCOMP   34
   2.3. DNA template requirements   35
3. Recommended protocols   35
   3.1. Downstream applications   40
4. Troubleshooting   44
5. References   45

Chapter 5
PRSG, a whole genome amplification method based on adaptor-ligation PCR of randomly sheared genomic DNA
Hiroki Sasaki and Kazuhiko Aoyagi
1. Introduction   47
2. Methods and approaches   48
3. Recommended protocols   48
   3.1. Downstream applications   55
4. Troubleshooting   57
5. References   57

Chapter 6
GenomePlex whole genome amplification
Simon Hughes, Gabrielle Sellick, Richard Coleman and John Langmore
1. Introduction   59
2. Methods and approaches   60
   2.1. GenomePlex WGA   60
3. Recommended protocols   62
   3.1. Downstream applications   68
4. Troubleshooting   74
5. References   75

Chapter 7
DNA linear amplification
Chih Long Liu, Bradley E. Bernstein and Stuart L. Schreiber
1. Introduction   77
2. Methods and approaches   79

2.1. DNA linear amplification   79
2.2. General strategy   79
2.3. Considerations for the starting dsDNA template   79
2.4. Using this method for ChIP–chip experiments   81
2.5. Controls for new users   82
3. Recommended protocols   84
3.1. Expected yields   91
3.2. Composition of the amplification product   92
3.3. Results   93
3.4. ChIP–chip results   93
4. Troubleshooting   94
4.1. Poor amplification yield   94
4.2. Poly(A) tracts in the amplified RNA   96
4.3. DsRNA formation in the amplified RNA product   97
5. References   98

Chapter 8
Multiple displacement amplification of genomic DNA
Roger Lasken

1. Introduction   99
1.1. The challenge of amplifying an entire genome   100
1.2. Reaction mechanisms and enzymology of MDA   102
2. Methods and approaches   106
2.1. Source of DNA template   106
2.2. Amount of biological specimen required as DNA template source   108
2.3. Source of MDA reagents   108
3. Recommended protocols   108
3.1. MDA reaction   108
3.2. Expected yield of amplified DNA and storage   110
3.3. Characteristics of amplified DNA and reaction optimization   110
4. Troubleshooting   117
5. References   117

Chapter 9
Multiple displacement amplification from single bacterial cells
Roger Lasken, Arumugham Raghunathan, Thomas Kvist, Thomas Ishøy,
Peter Westermann, Birgitte K. Ahring and Robert Boissy

1. Introduction   119
2. Methods and approaches   120
2.1. Isolation of single cells by serial dilution   120
2.2. Isolation of single cells by flow cytometry or micromanipulation 121
3. Recommended protocols   121
3.1. DNA sequencing with amplified DNA   123
3.2. Characteristics of DNA amplified by MDA from single cells   126
3.3. Isolation of single cells by micromanipulation and genomic DNA
     amplification by MDA   131

    3.4. Example of single-cell isolation and MDA   136

    3.5. Bioinformatic aspects of single-cell MDA for genome sequencing
        and assembly   141

  4. References   146

Chapter 10

Genome amplification tolerant to sample degradation: application to formalin-fixed, paraffin-embedded specimens

G. Mike Makrigiorgos

1. Introduction   149

2. Methods and approaches   150

    2.1. Principles of RCA–RCA   150

    2.2. Advantages of RCA–RCA   150

3. Recommended protocols   152

    3.1. Examples of results   154

4. Troubleshooting   159

5. References   160

Chapter 11

Pre-implantation genetic diagnosis using whole genome amplification

Alan H. Handyside, Mark D. Robinson and Francesco Fiorentino

1. Introduction   163

2. Methods and approaches   164

    2.1. PGD using WGA   164

3. Recommended protocols   169

    3.1. Downstream applications   170

4. References   183

Appendix 1

List of suppliers

List of suppliers   185

Index

Index   189

Contributors

Ahring, Birgitte K. Biocentrum-DTU, The Technical University of Denmark, Denmark. E-mail: bka@biocentrum.dtu.dk

Aoyagi, Kazuhiko Genetics Division, National Cancer Center Research Institute, 1-1, Tsukiji 5-chome, Chuo-ku, Tokyo 104-0045, Japan.
E-mail: kaaoyagi@gan2.res.ncc.go.jp

Arneson, Nona C.R. Ontario Cancer Institute, Division of Applied Molecular Oncology, University Health Network, 610 University Avenue, Room 10-717, Toronto, Ontario, Canada M5G 2M9. E-mail: nona.arneson@gmail.com

Bannai, Makoto Biomedical Business R&D Department, Olympus Corporation, 2-3 Kuboyama-cho, Hachioji-shi, Tokyo 192-8512, Japan.
E-mail: m_bannai@ot.olympus.co.jp

Bernstein, Bradley E. Department of Chemistry and Chemical Biology, Harvard University, 12 Oxford Street, Cambridge, MA 02138, USA.
E-mail: bbernstein@partners.org

Boissy, Robert Center for Genomic Sciences, Allegheny-Singer Research Institute, West Penn Allegheny Health System, 320 North East Avenue, Pittsburgh, PA 15212-4772, USA. E-mail: RBOISSY@wpahs.org

Coleman, Richard Institute of Cancer Research, Section of Cancer Genetics, 15 Cotswold Road, Surrey SM2 5NG, UK. E-mail: richard.coleman@icr.ac.uk

Dietmaier, Wolfgang Institute of Pathology, University of Regensburg, Franz-Josef-Strauss-Allee 11, 93053 Regensburg, Germany.
E-mail: wolfgang.dietmaier@klinik.uni-regensburg.de

Done, Susan J. Ontario Cancer Institute, Division of Applied Molecular Oncology, University Health Network, 610 University Avenue, Room 10-717, Toronto, Ontario, Canada M5G 2M9. E-mail: sdone@uhnres.utoronto.ca

Fiorentino, Francesco GENOMA, Molecular Genetics Labortory, Embryogen, Centre for Preimplantation Genetic Diagnosis, via Po, 102 00198 Rome, Italy. E-mail: fiorentino@laboratoriogenoma.it

Ghazani, Arezou A. Department of Laboratory Medicine and Pathobiology, University of Toronto, 610 University Avenue, Room 10-717, Toronto, Ontario, Canada M5G 2M9. E-mail: arezou.ghazani@utoronto.ca

Handyside, Alan H. London Bridge Fertility, Gynaecology and Genetics Centre, One St Thomas Street, London Bridge, London SE1 9RY, UK. E-mail: ahandyside@thebridgecentre.co.uk

Hughes, Simon Institute of Cancer Research, Section of Cancer Genetics, 15 Cotswold Road, Surrey SM2 5NG, UK. *Present address:* Tumour Biology Laboratory, John Vane Science Centre, Cancer Research UK Clinical Centre, Queen Mary's School of Medicine and Dentistry, Charterhouse Square, London EC1M 6BQ, UK. E-mail: simon.hughes@cancer.org.uk

Ishøy, Thomas Center for Genomic Sciences, Allegheny-Singer Research Institute, West Penn Allegheny Health System, 320 North East Avenue, Pittsburgh, PA 15212-4772, USA. E-mail: TISHOEY@wpahs.org

Kvist, Thomas Biocentrum-DTU, The Technical University of Denmark, Denmark. E-mail: tkv@biocentrum.dtu.dk

Lage, José M. Yale University School of Medicine, 310 Cedar Street, New Haven, CT 06520, USA. E-mail: jose.lage@yale.edu

Langmore, John Rubicon Genomics, 4370 Varsity Drive, Suite G, Ann Arbor, MI 48108, USA. E-mail: nagel@rubicongenomics.com

Lasken, Roger Center for Genomic Sciences, Allegheny-Singer Research Institute, West Penn Allegheny Health System, 320 North East Avenue, Pittsburgh, PA 15212-4772, USA. E-mail: rlasken2000@yahoo.com

Lizardi, Paul M. Yale University School of Medicine, 310 Cedar Street, New Haven, CT 06520, USA. E-mail: paul.lizardi@yale.edu

Liu, Chih Long Department of Chemistry and Chemical Biology, Harvard University, 12 Oxford Street, Cambridge, MA 02138, USA. E-mail: chihlongliu@gmail.com

Makrigiorgos, G. Mike Dana-Farber/Brigham and Women's Cancer Center, Level L2, Radiation Therapy, 75 Francis Street, Boston, MA 02115, USA. E-mail: mmakrigiorgos@LROC.harvard.edu

Raghunathan, Arumugham Qiagen Inc., 27220 Turnberry Lane, Valencia, CA 91354, USA. E-mail: Arumugham.Raghunathan@qiagen.com

Robinson, Mark D. Leeds Pre-implantation Genetic Diagnosis Centre, Assisted Conception Unit, Leeds General Infirmary, Leeds LS1 3EX, UK.
E-mail: m.d.robinson@leeds.ac.uk

Sasaki, Hiroki Genetics Division, National Cancer Center Research Institute, 1-1, Tsukiji 5-chome, Chuo-ku, Tokyo 104-0045, Japan.
E-mail: hksasaki@gan2.res.ncc.go.jp

Schreiber, Stuart L. Department of Chemistry and Chemical Biology, Harvard University, 12 Oxford Street, Cambridge, MA 02138, USA.
E-mail: stuart_schreiber@harvard.edu

Sellick, Gabrielle Institute of Cancer Research, Section of Cancer Genetics, 15 Cotswold Road, Surrey SM2 5NG, UK. E-mail: gabrielle.sellick@icr.ac.uk

Tokunaga, Katsushi Department of Human Genetics, Graduate School of Medicine, The University of Tokyo, 7-3-1 Hongo, Bunkyo-ku, Tokyo 113-0033, Japan. E-mail: tokunaga@m.u-tokyo.ac.jp

Westermann, Peter Biocentrum-DTU, The Technical University of Denmark, Denmark. E-mail: pw@biocentrum.dtu.dk

Wild, Peter J. Institute of Pathology, University of Regensburg, Franz-Josef-Strauss-Allee 11, 93053 Regensburg, Germany.
E-mail: peter-johannes.wild@klinik.uni-regensburg.de

Preface

Whole-genome amplification (WGA) methods began to appear in the literature in the early 1990s, with a variety of approaches emerging for the amplification of genome-representative DNA from limited sources for use in genetic testing. With the development of high-throughput genomics and DNA sequencing, the push for improved WGA technologies has accelerated. In this volume of the Scion book series *Methods Express*, we present a representative collection of the most commonly used WGA techniques. Detailed protocols are provided, along with sources for the required reagents and troubleshooting guides. The chapters are supported with references covering the development, testing, and validation of each method, along with examples of research applications. References of particular interest are also indicated with asterisks and guide the reader to papers of importance in understanding the uses, capabilities, and expected performance of each method.

Two categories of WGA technology have emerged. One group of methods uses variations of the polymerase chain reaction (PCR). Unlike conventional PCR, which targets a particular sequence, the application of PCR to WGA permits genome-wide amplification based on random or degenerate primers, or ligation of adaptors to genomic DNA fragments to provide a common PCR primer-binding site. Chapters 2 and 3 provide a description of degenerate-oligonucleotide-primed PCR and primer-extension pre-amplification, which were the first widely used PCR-based WGA techniques. Chapters 4–6 provide protocols for WGA using variations of the linker-adaptor technique, and describe how fragmentation of the genome by mechanical, enzymatic, or chemical methods, followed by ligation of adaptors and PCR, can allow the generation of microgram quantities of genome-representative DNA.

A second group of methods is based on isothermal DNA amplification. Multiple displacement amplification (MDA) uses φ29 DNA polymerase for WGA. Chapter 8 gives a thorough introduction to the basics of the MDA reaction, provides complete protocols, and explains its performance characteristics and use in downstream applications. Chapters 9–11 address specific applications for MDA from single bacterial cells, short DNA templates, and single human blastomeres, respectively. A method for isothermal linear amplification via RNA polymerase is also presented (Chapter 7). While yields can be lower than from exponential amplification-based methods, linear amplification can have the advantage of better conservation of quantitative information.

In some cases the reader may find it necessary to determine experimentally the method best suited for a particular study. However, a consensus has emerged on the general strengths and limitations of the different methods for some applications. The PCR-based methods have been widely used and have been validated, over a relatively long period, for amplifying DNA template for use in genetic testing, and for generating labelled probe for use in fluorescent *in situ* hybridization and other applications. Furthermore, the PCR-based methods have the advantage of efficiently amplifying very short DNA templates and thus may be preferred when using degraded DNAs, such as in forensic applications, and for WGA from paraffin-embedded tissues, which contain fragmented and cross-linked DNA.

The newer MDA reaction has taken us closer than ever before to fully replicating genomic DNA by a simple laboratory method. MDA generates DNA that is almost indistinguishable from the starting DNA template when used in many biotechnology applications. An estimated 99.8% of the genome is amplified with relatively even coverage of most sequences. ϕ29 DNA polymerase, with its $3'\rightarrow5'$ exonuclease proofreading activity, yields high-fidelity amplified DNA that gives a 99.9% accuracy in genotyping assays. In addition, MDA produces very large DNA products ranging up to hundreds of kilobases in length. MDA is clearly the method of choice for downstream applications requiring long DNA products, including library construction for DNA sequencing, cycle sequencing from PCR products, long PCR applications, and restriction fragment length polymorphism analysis.

Perhaps the ultimate challenge is to amplify genomic DNA from a single cell. The PCR-based methods and MDA have both made great strides in achieving this goal. Careful optimization is required and a greater percentage of reactions typically fail. Nevertheless, exciting new research strategies for single-cell analysis are now possible that promise to revolutionize many scientific fields. The chapters on genotyping embryos from single blastomeres and obtaining genomic sequence from single microbial cells introduce these approaches.

When considering WGA it is important to take into account the entire experimental process, from DNA source to required downstream applications, as many factors can affect DNA quality and thus have some bearing on the choice of WGA methodology. While we are not always able to give absolute recommendations for when to choose a particular WGA protocol, the reader is invited to participate in this exciting era of innovation and testing of the WGA methods. The next decade is expected to introduce many new research strategies taking advantage of WGA technology. The step-by-step methods provided in this volume will serve as an excellent guide for both experienced and novice researchers using WGA to tackle a variety of biological questions.

Simon Hughes and Roger Lasken
June 2005

Abbreviations

ADO	allele dropout
aRNA	antisense RNA
ATCC	American Type Culture Collection
BSA	bovine serum albumin
CGH	comparative genomic hybridization
ChIP	chromatin immunoprecipitation
CIP	calf intestinal alkaline phosphatase
DMSO	dimethyl sulfoxide
DOP–PCR	degenerate-oligonucleotide-primed PCR
DTT	dithiothreitol
FCS	fluorescence correlation spectroscopy
FFPE	formalin-fixed, paraffin-embedded
FISH	fluorescent *in situ* hybridization
GAPDH	glyceraldehyde phosphate dehydrogenase
HLA	human leukocyte antigen
I–PEP	improved primer-extension pre-amplification
IRS	interspersed repetitive sequence
IVT	*in vitro* transcription
LCM	laser-capture microdissection
LL–DOP–PCR	long products from low DNA quantities DOP–PCR
LM–PCR	ligation-mediated PCR
LOH	loss of heterozygosity
MDA	multiple displacement amplification
MFPE	methanol-fixed, paraffin-embedded
MSI	microsatellite instability
PBS	phosphate-buffered saline
PCR	polymerase chain reaction
PEP–PCR	primer-extension pre-amplification PCR
PGD	pre-implantation genetic diagnosis
PVP	polyvinylpyrrolidone
Q–PCR	quantitative PCR
RCA	rolling-circle amplification
RCA–RCA	restriction and circularization-aided rolling-circle amplification
RFLP	restriction fragment length polymorphism

R–PCR	random PCR
SCOMP	single-cell comparative genomic hybridization
SDS	sodium dodecyl sulfate
SNP	single nucleotide polymorphism
SSP–PCR	sequence-specific primer PCR
TAE	Tris-acetate-EDTA
TdT	terminal deoxynucleotidyl transferase
WGA	whole genome amplification

Color section

Chapter 4. Global amplification using SCOMP

a)

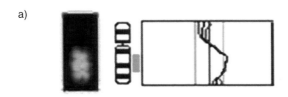

b)

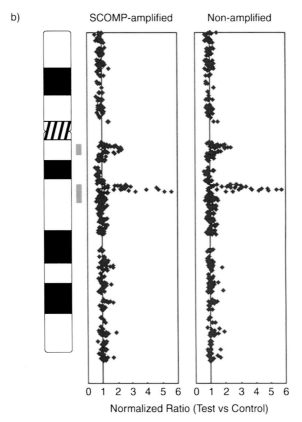

Figure 3. Chromosomal CGH (*a*) and array CGH (*b*) of a cell line with amplifications on the q arm of chromosome 17.
(*a*) Using SCOMP-amplified genomic DNA, chromosomal CGH shows clear hybridization signals and very little background. (*b*) Using SCOMP-amplified genomic DNA (left) and non-amplified genomic DNA (right), the array CGH profiles for chromosome 17 are almost indistinguishable and the regions of amplification are clearly detected using both templates. Chromosomal CGH shows a large amplification spanning 17q11–17q22, while array CGH is able to delineate two regions of amplification at 17q11.2 and 17q21.2–17q21.3. This example highlights the increased resolution and extended dynamic range of array CGH.

Chapter 8. Multiple displacement amplification of genomic DNA

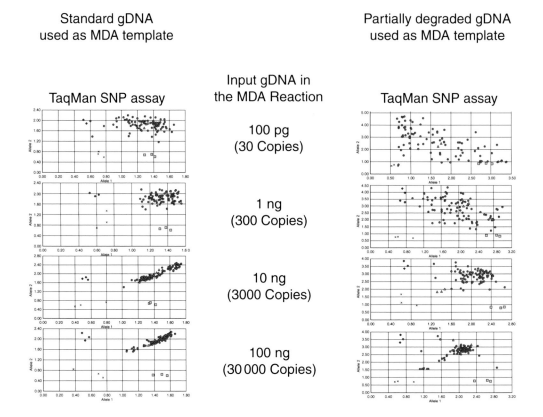

Figure 9. Effect of DNA template quality on MDA.
Purified genomic DNA (Standard gDNA) of an individual heterozygous for the test locus was amplified in 84 replicate MDA reactions. DNA that had been in long-term storage and was known to be of poor quality (Partially degraded gDNA) from a second heterozygous individual was also amplified in 84 replicate MDA reactions. The amplified DNA was then tested for genotyping by the TaqMan SNP assay (locus CYP2C19*2; ABI TaqMan Pre-Developed Allelic Discrimination Assay). Heterozygous genotypes should appear as a tight population along the diagonal of the graph. DNA was added as template for the MDA reactions in the amounts indicated. More of the degraded gDNA template was required (100 ng) than standard gDNA (10 ng) to generate amplified DNA optimal in the TaqMan assay.

Chapter 9. Multiple displacement amplification of single bacterial cells

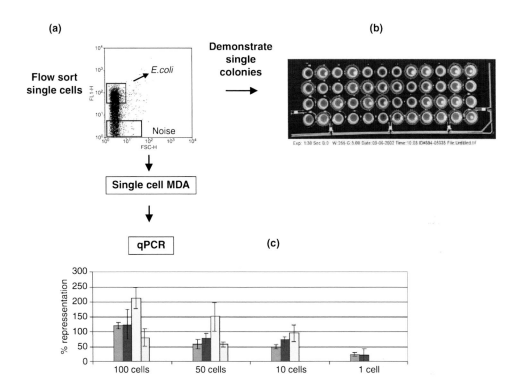

Figure 1. WGA from limited numbers of *E. coli* cells.
Single cells were collected by flow cytometry in microtiter plates. Approximately 100, 50, or 10 cells were also sorted into 96-well plates. Cells were lysed and subjected to MDA. The average locus representation was determined by quantitative PCR (qPCR) using TaqMan assays. Each bar represents the average of ten replicate MDA reactions for a test locus. Error bars are one standard deviation. (*a*) Isolation of single cells by flow cytometry. (*b*) Verification of the ability to capture single colony-forming units with flow cytometry by plating on agar in a microtiter plate. Twenty-six wells produced one colony (circled wells) and 22 wells had no colony-forming units. (*c*) TaqMan analysis following MDA from 100, 50, 10, and one cell as indicated.

Chapter 9. Multiple displacement amplification of single bacterial cells

(a)

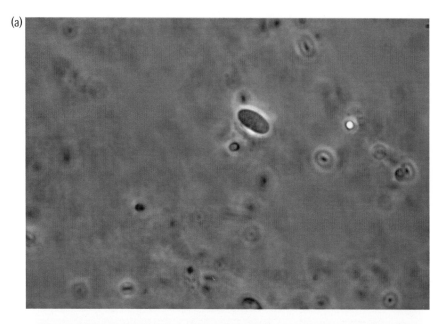

(b)

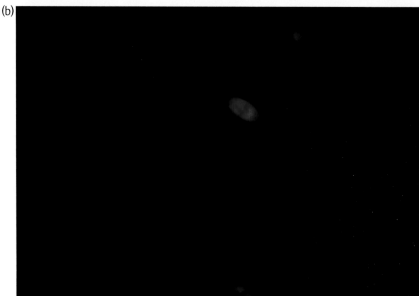

Figure 7. Labeling of cells for micromanipulation with FISH probes.
The figure shows two microscope pictures of the cell targeted for isolation. Both pictures
show the same section of the FISH-probed soil sample. The picture at the top is a bright
field image and the picture at the bottom shows the red emission from the Cy3-labeled
FISH-probe.

Chapter 11. Pre-implantation diagnosis using whole genome amplification

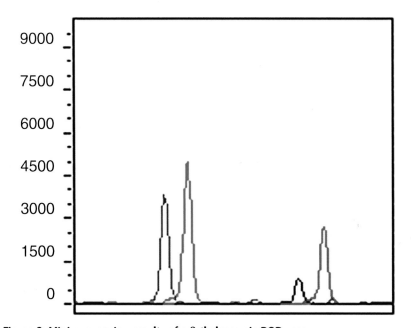

Figure 3. Mini-sequencing results of a β-thalassemia PGD case.
The mutations of interest were IVSI-110 G/A (left) and IVSI-6 T/C (right), analysed by a multiplex reaction. A compound heterozygote blastomere for the two mutations is shown. The multiplex single-base extension produces mini-sequencing products that differ significantly in size, so that it is possible to distinguish easily the two different mutation sites in a single capillary electrophoresis run. Color is assigned to individual ddNTPs as follows: A/green, C/black, G/blue and T/red. The mini-sequencing window (on the left, mutation IVSI-110 G/A) shows two different-colored peaks, one coming from the normal allele (blue peak) and the other from the mutated allele (green peak, mutant base A). On the right, mutation IVSI-6 T/C is shown as one red peak (normal allele, wild type base T) and one black peak (mutant base C).

Chapter 11. Pre-implantation diagnosis using whole genome amplification

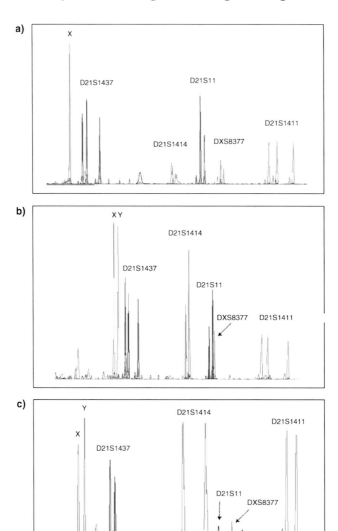

Figure 4. Multiplex quantitative fluorescent PCR assays with STR markers.
(a) Female embryo affected by trisomy 21, evidenced by a trisomic triallelic pattern (1:1:1) for D21S1437 and D21S1411, and a trisomic diallelic pattern (2:1) for D21S11 and D21S1414. (b) A male embryo affected by trisomy 21, evidenced by a trisomic triallelic pattern (1:1:1) for D21S1437 and D21S1411, and a trisomic diallelic pattern (2:1) for D21S11 and D21S1414. (c) A normal male embryo. For sex determination, amplification of a region of high homology between the X amelogenin and its Y pseudogene of the amelogenin X and Y sequences was carried out. Primers were designed at a location where there was 100% sequence homology between the X and Y amelogenin sequences. These primers flank a 6 bp deletion within intron 1 of the X homologue resulting in 104 bp and 110 bp PCR products from X and Y chromosomes, respectively. This size difference makes it possible to perform sex identification and is sufficient to allow easy discrimination between the two peaks (X and Y) after capillary electrophoresis. The STR marker DXS8377 was used to evaluate the copy number of chromosome X.

CHAPTER 1

Introduction to whole genome amplification

José M. Lage and Paul M. Lizardi

Yale University School of Medicine, New Haven, Connecticut, USA

1. INTRODUCTION

With the availability of the complete DNA sequence of many vertebrate and invertebrate genomes, functional genomic analysis has come of age. Exact sequence knowledge is uniquely enabling for a variety of powerful genetic analysis tools, most dramatically for microarray-based technologies. The expanding repertoire of analytical approaches available to define genetic alterations and unravel their relationship to human diseases has become a great driving force for discovery efforts in biomedical research. Nonetheless, in many research and clinical settings the availability of genomic DNA, in sufficient amounts, can be a major limitation with respect to the types of genomic analysis that can be considered. Such tests may include, amongst others, microsatellite analysis, mutation analysis, single nucleotide polymorphism (SNP) analysis, and comparative genomic hybridization (CGH) for determination of allele gains and losses, and in many instances the demand for DNA can rapidly outstrip the supply. In studies of cancer biology, where human biopsies acquired for diagnostic pathology represent a precious resource, the DNA source available from leftover biopsy samples often consists of only a few thousand cells and is frequently insufficient for performing the battery of genetic tests pertinent to the analysis of a cancer genome. Likewise, in studies where homogeneous tumor cell populations are generated by laser-capture microdissection, or unique subpopulations of immune cells are isolated by fluorescence-activated cell sorting, the total number of cells available for DNA extraction can be quite small. One of the most convenient sources of human DNA is a buccal swab, but, again, the amount of genetic material obtained in a single swab is rather limited. Obviously, this is a problem begging for a DNA amplification solution.

The development of the polymerase chain reaction (PCR) in the 1980s (1) rapidly led to the realization that it might be possible to amplify entire genomes.

Whole Genome Amplification: *Methods Express* (S. Hughes and R. Lasken, eds.)
© Scion Publishing Limited, 2005

However, whole genome amplification (WGA) is the antithesis of single-locus amplification by PCR, with WGA having the goal that all gene loci should be amplified with low bias. In due time the challenge was met, and during the last two decades a variety of methods have been developed that are capable of amplifying total genomic DNA. This volume of the *Methods Express* series comprises a good sampling of the currently available repertoire of WGA techniques. In this introduction we discuss a number of interesting mechanistic considerations relevant to different WGA approaches. We also dwell briefly on the challenges of single-cell WGA and comment on existing opportunities for the future development of even more powerful forms of WGA.

2. METHODS AND APPROACHES

The earliest WGA approaches employed primers complementary to abundant genomic DNA repeats, or random or partially degenerate primers, combined with specially modified thermocycling protocols (2–6). Such methods yielded a few hundred copies of the genome, and the size of the DNA product ranged from 200 to 3000 bases. More elaborate methods were soon developed, capable of generating higher amplification yields, by virtue of the ligation of oligonucleotide adaptors to genomic fragments, thereby enabling more efficient priming during PCR cycling (7–9). During the last 10 years, we have witnessed the development of nonPCR-based methods, some of which rely on DNA strand displacement and others on linear, rather than exponential, DNA amplification. Some of these new approaches are quite simple and do not require thermocycling. For a thorough overview of the literature on WGA technology, the reader is referred to a recent review by Hughes *et al.* (10).

The earliest WGA method employing PCR, interspersed repetitive sequence PCR (IRS–PCR) (2), used primers designed to anneal to specific repetitive sequences (Alu repeats) present in the genome. Unfortunately, it relied on the repeats being evenly distributed throughout the genome; in addition, the primers needed to be close enough to one another to permit the generation of a PCR product. As a result it introduced bias by preferentially amplifying regions of the genome rich in Alu repeats. The next PCR-based WGA methods used random or partially degenerate primers, especially degenerate-oligonucleotide-primed PCR (DOP–PCR; see Chapter 2) (3) and primer-extension pre-amplification PCR (PEP–PCR; see Chapter 3) (4). These methods have been widely used and do not suffer from the limitations of IRS–PCR. Since their conception, improved variants such as improved PEP (I–PEP) (11) and long products from low DNA quantities DOP–PCR (LL–DOP–PCR) (12) have led to more robust and less biased representations of the genome. A fundamental difference between DOP–PCR and PEP–PCR is that in the former the potential priming sites in genomic DNA are a limited subset (probably of the order of 10^6 sites per genome), while in the latter, which uses random primers, the number of potential priming sites is orders of magnitude larger. Thus, on theoretical grounds, PEP or I–PEP is likely to generate less-biased representations, a view that is supported by a number of reports in the literature (for review, see 10). To overcome the limitation of available priming sites, alternative approaches based on DNA fragmentation, followed by either the ligation of oligonucleotide adaptors (see Chapters 4–6) or the addition of poly(dT) tails by terminal deoxynucleotidyl transferase (see Chapter 7) have been developed. These have the advantage that only one or two primer sequences are needed, which results in a more efficient amplification and larger yields. DNA fragmentation can be accomplished using sequence-specific restriction endonucleases, random endonucleolytic cleavage, or random shear. In some instances, the sample is already fragmented or degraded, as is the case in many formalin-fixed, paraffin-embedded specimens. Random cleavage (or shear) has the theoretical advantage that the different populations of DNA amplicons comprise a random, partially overlapping distribution of sequences, and hence the averaged amplification yield for different genomic loci is less likely to display sequence-dependent fluctuations. In other words, the amplicons resulting from restriction endonuclease cleavage are highly discrete, while the amplicons generated by random breaks represent a continuum of fragmented molecules, more likely to provide equal representation of all sequences as these are amplified in different contexts. Liu *et al.* (13; see Chapter 7) have amplified DNA obtained after chromatin immunoprecipitation (ChIP) using a linear amplification method based on DNA tailing and one or several rounds of transcription by T7 RNA polymerase. Linear amplification is insensitive to the content of repeats in the template DNA, which in the case of PCR-based methods can reduce the relative representation of abundant sequences due to competition between primer annealing and reassociation of complementary strands (14). An advantage of using linear amplification methodology is the lack of dynamic range compression in ChIP–chip experiments compared with PCR-based approaches (see Chapter 7).

A limitation of any WGA method is that the likelihood of representational bias increases as the number of initial template molecules approaches very small numbers (less than ten genomes). At such low input levels, the possibility of random drop-off of individual chromosomal loci is always a possibility. Thus, although single-cell comparative genomic hybridization (SCOMP; see Chapter 4) analyses were reported as early as 1999 (9), such experiments are challenging and users must be aware of the potential for loss of gene loci, unless experimental procedures are carefully controlled. Nevertheless, rapid advances are being made in the use of WGA for pre-implantation diagnosis from single blastomeres isolated from human embryos (see Chapter 11) and for genomic sequencing from single bacterial cells for which the lack of culture methods has, until now, limited studies of the vast numbers of undescribed microbes (see Chapter 9).

A mathematical model for two different PCR-based WGA reactions (PEP–PCR and tagged random primer PCR) was developed by Sun *et al.* (15) with the objective of exploring predictions of target yield and coverage. One of the interesting insights generated by their mathematical model was that the use of a DNA polymerase with high processivity would lead to increased amplification efficiency and locus coverage. Familiarity with the work of Arnheim and his collaborators influenced our own thinking in the late 1990s about the utilization of strand-displacing DNA polymerases for isothermal DNA amplification, which had been successfully adapted for the amplification of small DNA loci by Terry Walker (16). Our group at Yale and the group of Roger Lasken became interested in WGA as we began to evaluate *in vitro* DNA amplification reactions catalyzed by strand-displacing DNA polymerases, in particular rolling-circle amplification, or RCA (17–20). We decided against the use of DNA holoenzymes, which are highly efficient in strand displacement, because the requirement for multiple enzyme subunits made them less attractive for developing robust *in vitro* applications. We evaluated a number of single-subunit DNA polymerases known to have a strong DNA strand-displacement activity, as well as polymerases with a weak strand-displacement activity that could be augmented by the addition of single-strand-binding proteins such as phage T4 gene 32 protein. Among these enzymes, the most interesting turned out to be the DNA polymerase from phage φ29, a B family polymerase that initiates replication using a protein as a primer, which attaches the first nucleotide of the phage genome to the hydroxyl of a specific serine of the priming protein. Pioneering studies by the group of Margarita Salas (21) had shown that this enzyme is capable of highly processive DNA replication in the absence of any accessory protein and is able to produce strand displacement coupled to the polymerization process, generating DNA strands in excess of 70 kb. Our early work on RCA at Yale was made possible by generous gifts of this precious reagent by Dr. Salas. Recently, the crystal structure of φ29 DNA polymerase has been determined at 2.2 Å resolution (22) and this structure has generated fascinating insights regarding its extraordinary processivity and strand-displacement activities. Homology modeling suggests that downstream template DNA passes through a tunnel prior to entering the polymerase active site. This tunnel is too small to accommodate dsDNA and requires the separation of template and nontemplate strands. Thus, the high processivity of φ29 DNA polymerase may be explained by its topological encirclement of both the downstream template and the upstream duplex DNA.

Strand-displacing enzymes such as φ29 DNA polymerase and *Bst* DNA polymerase are capable of utilizing synthetic oligonucleotides as primers. The ability of the enzymes to displace dsDNA gives rise to branching of ssDNA in a mechanism called hyper-branching (19). In the presence of random primers, multiple priming events occur (see *Fig. 1*). Protection of the primers with phosphorothioate linkages at the 3′ terminus prevented their degradation by the 3′→5′ exonuclease activity of φ29 DNA polymerase and enabled a large increase in the level of amplification (23), with plasmids being amplified more than 10 000-fold. The use of the reaction, termed multiple-displacement amplification (MDA, US Patent No. 6124120) to amplify the entire human genome was first described by Dean *et al.* (24) and subsequently by Lage *et al.* (25). Using a microarray that contained 4600 cDNA probes, Lage *et al.* observed that the DNA generated by MDA contained approximately equal copy numbers of all probed gene loci. In those studies, the sequence representation in the amplified DNA was somewhat better for reactions using *Bst* DNA polymerase compared with φ29 DNA polymerase. A drawback of using *Bst* DNA polymerase, however, is the low fidelity of this enzyme. It has been documented that φ29 DNA polymerase has associated

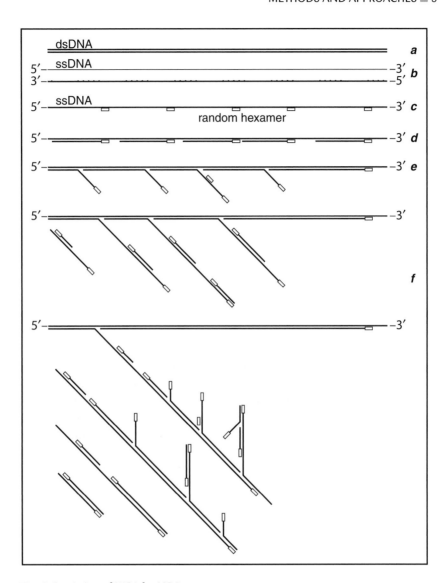

Fig. 1. Depiction of WGA by MDA.
Genomic dsDNA (*a*) is denatured by alkali or heat treatment, generating ssDNA (*b*). For simplicity, only a positive strand is shown, but amplification takes place in both strands simultaneously. Random hexamers (white rectangles) attach to the genomic strands at multiple positions (*c*). Primers are extended by a DNA polymerase with strand-displacement activity (*d*), which is able to displace strands from the double-stranded domains (*e*). The displaced strands are targeted by more random hexamers (*e*, *f*) whose extension generates additional branches, leading to the generation of hyper-branched structures.

proofreading activity (26) that contributes to high fidelity in DNA replication, as well as low accumulation of mutations during amplification by MDA (27). Also, the high concentration of φ29 DNA polymerase (24) results in an amplification bias of less than sixfold, and in nicely balanced representations of the entire genome, based on quantitative PCR of 47 loci, one on the p and q arm of each human chromosome (28). One exception is the apparent under-representation of some repetitive DNA sequences in the amplified DNA. It has been suggested (24) that this may be caused by depletion of primers corresponding to highly repeated DNA. However, straightforward calculations show that the concentration of specific hexamer sequences is in large excess after a 2000-fold amplification reaction initiated by 12 ng of DNA, relative to the available template concentration of any specific primable hexamer of the Alu sequence, the most abundant repeat in the human genome. Thus, the relatively lower representation of DNA repeats in amplified DNA is likely to be caused by other molecular mechanisms that we do not yet understand.

Recently, our group at Yale has modeled the process of DNA amplification driven by random primers in the presence of a strand-displacing DNA polymerase. The simulations, which will be reported in detail elsewhere, show the precise manner in which the yield of amplified DNA is influenced by DNA template size as well as by the stochasticity of priming events, in such a way that every single molecule of initial DNA template gives rise to a different, uniquely skewed genealogy of offspring molecules. Nonetheless, in a population containing hundreds of initial template molecules, the stochastic skew effects balance each other out. The simulations predict that, as the reaction proceeds, the fraction of ssDNA decreases and eventually stabilizes at around 10–20% of the total sequence. We believe that the computer simulations can help to shed some light on why the MDA reaction is self-limiting as currently implemented. As shown in *Fig. 2*, the computer simulations accurately predict that the yield of DNA should be influenced by the size of template DNA molecules. Additionally, the simulations show that, for any short DNA subsegment, the amplification yield of the subsegment is strongly dependent on the distance from the terminus of the DNA template. Of course, total genomic DNA as isolated in the laboratory tends to become fragmented at random by shear forces, resulting in a distribution of overlapping fragments. The stochasticity of genomic DNA fragmentation largely cancels out the potential problem of differential yields of DNA sequences relative to DNA termini. A special situation exists for molecules with fixed ends, such as those containing the terminus of a chromosome, whereby many yeast sequences located close to telomeres are amplified less (25). The poor performance of MDA when using genomic DNA that has been fragmented to a size of less than 5 kb makes it impractical to apply the standard MDA technique to DNA extracted from archival pathology tissue samples. A recent improvement reported by Wang *et al.* (29) utilizes restriction enzyme digestion of archival DNA, followed by ligation of the fragmented DNA to form circular templates for WGA, thereby effectively transforming the reaction into RCA-mediated WGA (restriction and circularization-aided RCA (RCA–RCA); see Chapter 10). Wang *et al.* reported that this methodological improvement resulted in a good amplification yield and

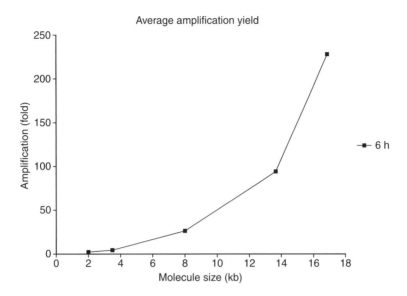

Fig. 2. Yields of simulated MDA reactions for DNA fragments of different lengths.
Computer simulations of the MDA reaction were run using the agent-based modeling tool
REPAST (Social Sciences Computing Services, University of Chicago, USA) for five different
fragment sizes (2.0, 3.5, 8.0, 13.6 and 16.8 kb). Values for the 2.0, 3.5 and 8.0 kb
fragments correspond to an average of 20 simulated runs. Values for the 13.6 and
16.8 kb fragments correspond to an average of six and two simulated runs, respectively.
The simulations show the effect of strand length on the total DNA amplification yield after
a 6 h reaction. Shorter DNA molecules are amplified much less than longer ones,
illustrating the profound effect of DNA size in these reactions.

improved homogeneity of locus representation relative to MDA. This procedure is
also applicable to the amplification of cDNA.

In this volume of *Methods Express*, the contributing authors present
optimized methods and exemplary applications for DOP–PCR (Chapter 2, Bannai
& Tokunaga) as well as I–PEP–PCR (Chapter 3, Wild & Dietmaier). The chapter by
Arneson *et al.* describes methods and applications of SCOMP (Chapter 4) for the
genetic analysis of single cells. Sasaki & Aoyagi describe a WGA method using
adaptor-ligation PCR of randomly sheared genomic DNA, or PRSG (Chapter 5), and
Hughes *et al.* describe GenomePlex WGA (Chapter 6), a method that is also based
on the use of random fragmentation of DNA. DNA linear amplification is the
subject of the chapter by Liu *et al.* (Chapter 7). They illustrate the adaptation of
this powerful method to WGA of chromatin-immunoprecipitated DNA. Roger
Lasken presents chapters on MDA methods (Chapter 8) and its use for WGA from
single bacterial cells (Chapter 9, Lasken *et al.*). Mike Makrigiorgos presents
RCA–RCA, an MDA method based on strand-displacement amplification of
circularized DNA (Chapter 10). Finally, Alan Handyside describes the application of
MDA to the study of single blastomeres (Chapter 11, Handyside *et al.*). The various

chapters will expose the reader to applications of WGA in the fields of SNP analysis, microsatellite analysis, and CGH, among others.

As WGA technology has matured and become more robust, some of the WGA methods have become available as commercial kits. Nonetheless, important challenges remain for the development of even more powerful WGA technology. For example, it would be highly attractive to develop a WGA method capable of preserving faithfully all the specific DNA methylation marks present in the genome of any given cell type. Thus, future developments in WGA may profit from the utilization of multiple protein subunits and multi-enzyme complexes, including the methylation mark maintenance machinery.

Acknowledgements

We thank Sebastian Szpakowski, Jill Rubinstein, and János Lobb for their work in the development of MDA computer simulations. This work has been supported by the National Cancer Institute (grants CA81671 and CA099135 to P.M.L.).

3. REFERENCES

1. Saiki RK, Bugawan TL, Horn GT, Mullis KB & Erlich HA (1986) *Nature*, **324**, 163–166.
2. Nelson DL, Ledbetter SA, Corbo L, *et al.* (1989) *Proc. Natl. Acad. Sci. U. S. A.* **86**, 6686–6690.
3. Telenius H, Carter NP, Bebb CE, Nordenskjold M, Ponder BA & Tunnacliffe A (1992) *Genomics*, **13**, 718–725.
4. Zhang L, Cui X, Schmitt K, Hubert R, Navidi W & Arnheim N (1992) *Proc. Natl. Acad. Sci. U. S. A.* **89**, 5847–5851.
5. Grothues D, Cantor CR & Smith CL (1993) *Nucleic Acids Res.* **21**, 1321–1322.
6. Cheung VG & Nelson SF (1996) *Proc. Natl. Acad. Sci. U. S. A.* **93**, 14676–14679.
7. Lüdecke HJ, Senger G, Claussen U & Horsthemke B (1989) *Nature*, **338**, 348–350.
8. Saunders RD, Glover DM, Ashburner M, *et al.* (1989) *Nucleic Acids Res.* **17**, 9027–9037.
9. Klein CA, Schmidt-Kittler O, Schardt JA, Pantel K, Speicher MR & Riethmuller G (1999) *Proc. Natl. Acad. Sci. U. S. A.* **96**, 4494–4499.
★★ 10. Hughes S, Arneson N, Done S & Squire J (2005) *Prog. Biophys. Mol. Biol.* **88**, 173–189. – *A thorough overview of WGA technology.*
11. Dietmaier W, Hartmann A, Wallinger S, *et al.* (1999) *Am. J. Pathol.* **154**, 83–95.
12. Kittler R, Stoneking M & Kayser M (2002) *Anal. Biochem.* **300**, 237–244.
13. Liu CL, Schreiber SL & Bernstein BE (2003) *BMC Genomics*, **4**, 19.
14. Mathieu-Daudé F, Welsh J, Vogt T & McClelland M (1996) *Nucleic Acids Res.* **24**, 2080–2086.
15. Sun F, Arnheim N & Waterman MS (1995) *Nucleic Acids Res.* **23**, 3034–3040.
16. Walker GT, Fraiser MS, Schram JL, Little MC, Nadeau JG & Malinowski DP (1992) *Nucleic Acids Res.* **20**, 1691–1696.
17. Fire A & Xu SQ (1995) *Proc. Natl. Acad. Sci. U. S. A.* **92**, 4641–4645.
18. Kool ET (1996) *Annu. Rev. Biophys. Biomol. Struct.* **25**, 1–28.
19. Lizardi PM, Huang X, Zhu Z, Bray-Ward P, Thomas DC & Ward DC (1998) *Nat. Genet.* **19**, 225–232.
20. Baner J, Nilsson M, Mendel-Hartvig M & Landegren U (1998) *Nucleic Acids Res.* **26**, 5073–5078.
21. Blanco L, Bernad A, Lazaro JM, Martin G, Garmendia C & Salas M (1989) *J. Biol. Chem.* **264**, 8935–8940.

22. Kamtekar S, Berman AJ, Wang J, *et al.* (2004) *Mol. Cell*, **16**, 609–618. Erratum **16**, 1035–1036.
23. Dean FB, Nelson JR, Giesler TL & Lasken RS (2001) *Genome Res.* **11**, 1095–1099.
24. Dean FB, Hosono S, Fang L, *et al.* (2002) *Proc. Natl. Acad. Sci. U. S. A.* **99**, 5261–5266.
25. Lage JM, Leamon JH, Pejovic T, *et al.* (2003) *Genome Res.* **13**, 294–307.
26. Esteban JA, Salas M, & Blanco L (1993) *J. Biol. Chem.* **268**, 2719–2726.
27. Nelson JR, Cai YC, Giesler TL, *et al.* (2002) *Biotechniques*, **Suppl.**, 44–47.
28. Hosono S, Faruqi AF, Dean FB, *et al.* (2003) *Genome Res.* **13**, 954–964.
29. Wang G, Maher E, Brennan C, *et al.* (2004) *Genome Res.* **14**, 2357–2366.

CHAPTER 2

Single nucleotide polymorphism typing using degenerate-oligonucleotide-primed PCR-amplified products

Makoto Bannai[1] and Katsushi Tokunaga[2]

[1]Biomedical Business R&D Department, Olympus Corporation, 2-3 Kuboyama-cho, Hachioji-shi, Tokyo 192-8512, Japan; [2]Department of Human Genetics, Graduate School of Medicine, The University of Tokyo, 7-3-1 Hongo, Bunkyo-ku, Tokyo 113-0033, Japan

1. INTRODUCTION

WGA is a valuable technique for amplifying a limited amount of DNA in a sequence-independent fashion. WGA methods have been adopted for minimization of the amount of genomic DNA needed for a number of biological assays including large-scale typing of single nucleotide polymorphisms (SNPs), microsatellite genotyping, and comparative genomic hybridization (CGH).

WGA by degenerate-oligonucleotide-primed PCR (DOP–PCR) was first described by Telenius *et al.* (1) and allows complete genome coverage in a single reaction. In contrast to the pairs of target-specific primer sequences used in traditional PCR, only a single primer, which has defined sequences at its 5′ (containing a *Xho*I restriction site, highlighted in bold) and 3′ ends and a random hexamer sequence between them (5′-CCGA**CTCGAG**NNNNNNATGTGG-3′), is used for DOP–PCR. Compared with completely degenerate primers, such as those used for primer-extension pre-amplification PCR (PEP–PCR) (5′-NNNNNNNNNNNNNNN-3′), the primer for DOP–PCR is relatively specific (2).

DOP–PCR is comprised of two different cycling stages, low stringency and high stringency. At low stringency, the 3′ end of the primer binds at sites in the genome complementary to the 6 bp well-defined sequence (approximately 10^6 sites in the human genome). The adjacent random hexamer sequence, which displays all possible combinations of the nucleotides A, G, C, and T, then enables efficient primer annealing and the start of the DOP–PCR-based WGA reaction.

Whole Genome Amplification: *Methods Express* (S. Hughes and R. Lasken, eds.)

Since its conception, several modifications of the basic DOP–PCR protocol have been devised with the purpose of lowering the required amount of starting template (3) and increasing yield (4), fidelity, and fragment length (5), in order to provide better coverage of the genome. However, all have used the same basic methodology.

2. METHODS AND APPROACHES

2.1. Methodology of DOP–PCR

Within a single PCR tube, low-temperature annealing and extension in the first five to eight cycles of DOP–PCR occurs at many binding sites in the genome (see *Fig. 1a* and *Protocol 2*, Stage 1 – low stringency) and tags these sequences with the DOP primer. Thereafter, the annealing temperature of the PCR (>25 cycles) is increased to allow more specific priming and amplification of the tagged sequence (see *Fig. 1a* and *Protocol 2*, Stage 2 – high stringency). DOP–PCR amplification ideally results in a smear of DNA fragments (200–1000 bp) that are visible on an agarose gel stained with ethidium bromide (see *Fig. 1b*).

2.2. Applications of DOP–PCR

DOP–PCR is often used as the first step in *in situ* hybridization for flow-sorted (1, 6) or microdissected (7) chromosomes and for CGH (7, 8). This approach has been successfully modified and applied to genomic DNA for genotyping of microsatellites (9) and for typing of SNPs (10–12). In this chapter, we describe sequence-specific primer PCR (SSP–PCR) followed by fluorescence correlation spectroscopy (FCS) as a method for applying DOP–PCR to SNP typing (11).

3. RECOMMENDED PROTOCOLS

Although we have named specific suppliers for the reagents and equipment used in this chapter, other manufacturers' products are likely to generate similar results. However, it is up to the user to test this.

(a)

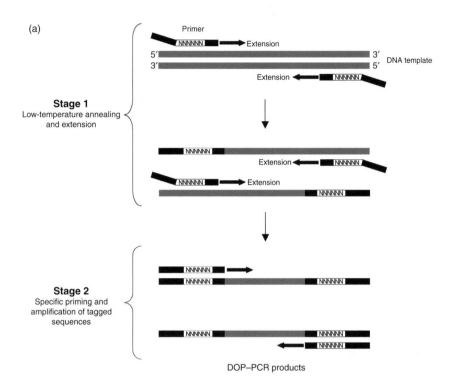

Stage 1
Low-temperature annealing
and extension

Stage 2
Specific priming and
amplification of tagged
sequences

DOP–PCR products

(b)

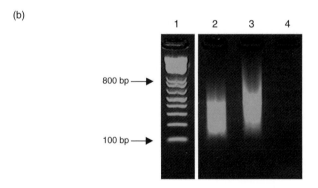

Figure 1. DOP–PCR.
(*a*) Graphical representation of the steps involved in DOP–PCR. Low-temperature
annealing and extension in the first five to eight cycles of DOP–PCR occurs at many
binding sites in the genome (Stage 1) and tags these sequences with the DOP primer.
Thereafter, the annealing temperature of the PCR (>25 cycles) is increased to allow more
specific priming and amplification of the tagged sequence (Stage 2). (*b*) Agarose gel
stained with ethidium bromide displaying the smear of DNA fragments typically obtained
from DOP–PCR WGA. (*Lane 1*) 100 bp ladder; (*lane 2*) DOP–PCR product obtained from
DNA extracted from formalin-fixed, paraffin-embedded tissue; (*lane 3*) DOP–PCR product
obtained from DNA extracted from fresh tissue; (*lane 4*) negative control.

Protocol 1

Genomic DNA extraction and quantification

Equipment and Reagents
- QIAamp DNA Blood Midi Kit (Qiagen)
- PicoGreen dsDNA quantitation reagent (Molecular Probes)
- 10 mM Tris-HCl (pH 8.0) containing 0.1 mM EDTA
- Fluorescence-based microplate reader

Method
1. Isolate genomic DNA from peripheral blood using the QIAamp DNA Blood Midi Kit according to the manufacturer's instructions.

2. Measure the genomic DNA concentration using the PicoGreen dsDNA quantitation reagent according to the manufacturer's instructions.

3. Dilute the DNA samples to 10 ng/μl with 10 mM Tris-HCl (pH 8.0) containing 0.1 mM EDTA.

DNA yields of approximately 4–6 μg are typically obtained in a 100 μl reaction. The diluted DOP–PCR product (approximately 10–15 ng/μl) can be used as template in a subsequent PCR to generate fragments including SNP sites. The PCR products can then be sequenced.

SSP–PCR followed by FCS is applied for high-throughput SNP typing and has been described previously (11). The first PCR is followed by SSP–PCR using the product from the first PCR as template. Allele-specific, semi-nested primers are used for the SSP–PCR. They differ in a single nucleotide at the 3′ end and are coupled to different fluorescence dyes – 6-carboxytetramethylrhodamine (TAMRA) or cyanine 5 (Cy5). Because the movement of DNA fragments in a solution depends on their size, primers (smaller molecules) move faster than SSP–PCR-amplified fragments (larger molecules) in the PCR product solution. When the narrow laser beam spots DNA fragments in the solution (in a 1 fl volume), the signals from the fluorescent-labeled molecules are detected by a highly sensitive spectrophotometer, allowing determination of the numbers and sizes of both primers and amplified fragments. The percentage of allele-specific amplified fragments relative to the total number of fluorescent-labeled molecules can then be determined using the single-molecule fluorescence detection system.

Protocol 2

DOP–PCR[a]

Equipment and Reagents

- 4 µM DOP primer (5′-CCGACTCGAGNNNNNNNATGTGG-3′) (Sigma Genosys)
- TaKaRa LA *Taq* polymerase (5 units/µl) and accompanying 2× GC buffer (Takara Bio)
- 400 µM dNTP mix (Takara Bio)
- Nuclease-free water (Sigma)
- 10 mM Tris-HCl (pH 8.0) containing 0.1 mM EDTA
- Thermal cycler (MJ Research)
- Agarose (Amersham Biosciences)
- Ethidium bromide (10 mg/ml) (Sigma)
- 1× TBE buffer (89 mM Tris; 89 mM boric acid; 2 mM EDTA)
- Electrophoresis apparatus
- Spectrophotometer

Method

1. Use 10 ng of genomic DNA as a template in a DOP–PCR mixture containing 1 µl of TaKaRa LA *Taq* polymerase, 50 µl of 2× GC buffer, 400 µM dNTPs, 4 µM DOP primer and nuclease-free water up to a final volume of 100 µl[b].

2. Perform DOP–PCR in a thermal cycler with an initial incubation of 93°C for 1 min[c], followed by eight cycles of 93°C for 1 min, 30°C for 1 min, and 72°C for 3 min (Stage 1)[d], and 28 cycles of 93°C for 1 min, 60°C for 1 min, and 72°C for 3 min (Stage 2)[e].

3. Run 5–10 µl of the DOP–PCR products including the negative control on a 1% agarose gel to assess fragment size and the success of the reaction[f].

4. Dilute the DOP–PCR products with four volumes of 10 mM Tris-HCl (pH 8.0) containing 0.1 mM EDTA and store at –20°C until use.

Notes

[a]The DOP–PCR method is as described previously (9) with slight modifications.

[b]It is important to include a negative control, which includes all of the reaction constituents with the exception of DNA.

[c]The initial denaturation for 8 min at 96°C, as suggested in (9), can be omitted with no effect on the efficiency of the DOP–PCR protocol, at least for the PCR targets tested in our work (11).

[d]Low-temperature annealing and extension, which occurs at several binding sites across the genome.

[e]Elevated annealing temperature, allowing more specific priming of the fragments tagged with the primer sequence.

[f]The negative control lane should not show any amplification. If it does, this suggests possible contamination and therefore reactions must be repeated. We would suggest using fresh reagents.

Protocol 3

Direct sequencing for SNP analysis

Equipment and Reagents

- AmpliTaq Gold DNA polymerase (5 units/μl) with the GeneAmp 10× PCR Gold Buffer (Applied Biosystems), or another comparable hot-start enzyme
- 500 nM PCR primers (Sigma Genosys)
- 25 mM MgCl₂ stock solution (Roche Diagnostics)
- 200 μM dNTP mix (Takara Bio)
- QIAquick PCR Purification Kit (Qiagen)
- Spectrophotometer
- BigDye Terminator Version 3.1 Cycle Sequencing Kit (Applied Biosystems)
- DNA sequencer (Applied Biosystems)
- Thermal cycler (MJ Research)
- SEQSCAPE Version 2.0 (Applied Biosystems)

Method

1. Prepare a 10 μl PCR mixture containing 1 μl of the diluted DOP–PCR product (approximately 10–15 ng of amplified DNA), 0.5 units of AmpliTaq Gold DNA polymerase, 200 μM dNTPs, 3.1 mM MgCl₂, and 500 nM of each primer.

2. PCR amplify using the following protocol: initial incubation at 95°C for 10 min; followed by 40 cycles of 95°C for 30 s, optimal annealing temperature for each primer pair[a] for 30 s, and 72°C for 1 min; and a final incubation at 72°C for 10 min.

3. Purify the PCR products using the QIAquick PCR Purification Kit, following the manufacturer's instructions.

4. Determine the yield of the PCR by measuring absorbance at 260 nm using a spectrophotometer.

5. Using the PCR products as templates, perform a cycle sequencing reaction using the BigDye Terminator Version 3.1 Kit according to the manufacturer's instructions[b].

6. Perform direct sequencing using a DNA sequencer and analyze the SNP types using SEQSCAPE Version 2.0 software[c].

Notes

[a]The primer pairs and optimal annealing temperatures used for these experiments are specific for the SNPs of interest and thus should be determined for each experiment.

[b]The quantity of PCR product used for sequencing varies depending on the size of the product. For 100–200 bp, use 1–3 ng of DNA; for 200–500 bp, use 3–10 ng of DNA; for 500–1000 bp, use 5–20 ng of DNA; for 1000–2000 bp, use 10–40 ng of DNA; for >2000 bp, use 20–50 ng of DNA (www.appliedbiosystems.com).

[c]Other instruments and software may also be used for the sequence analysis.

Protocol 4

High-throughput SNP typing[a]

Equipment and Reagents
- AmpliTaq DNA polymerase, Stoffel fragment (5 units/µl) with the 10× Stoffel buffer (Applied Biosystems)
- 20 nM TAMRA-labeled SSP–PCR primer (Sigma Genosys)
- 20 nM Cy5-labeled SSP–PCR primer (Sigma Genosys)
- 25 mM MgCl$_2$ stock solution (Roche Diagnostics)
- 200 µM dNTP mix (Takara Bio)
- 10 mM Tris-HCl (pH 8.0)
- 384-Well, hard-shell, thin-walled plates (MJ Research)
- 384-Well, glass-bottomed plates (Olympus Corporation)
- Single-molecule fluorescence detection (SMFD) system (Olympus Corporation)
- Thermal cycler (MJ Research)

Method
1. Using diluted DOP–PCR products as templates, perform PCR to amplify a fragment including an SNP site (see *Protocol 3*).

2. Perform SSP–PCR using the two competitive allele-specific primers in a 10 µl reaction containing 1× Stoffel buffer, 0.5 units of AmpliTaq DNA polymerase Stoffel fragment, 200 µM dNTPs, 2.5 mM MgCl$_2$, 20 nM of each primer, and 0.5 µl of the first PCR product as template.

3. PCR amplify in 384-well, hard-shell, thin-walled plates using the following protocol: initial incubation at 95°C for 2 min; followed by 40 cycles of ramping at 0.1°C/s to 95°C, 95°C for 30 s, optimal annealing temperature for each SNP for 30 s, and 72°C for 30 s; and a final incubation at 72°C for 10 min.

4. Transfer 4 µl of the SSP–PCR products into separate wells of a 384-well, glass-bottomed plate and dilute with 24 µl of 10 mM Tris-HCl (pH 8.0).

5. Analyze SNPs in the SSP–PCR products by FCS using the SMFD system[b]. Measure fluorescence at both 543 and 633 nm excitation wavelengths. Subject the mixture to FCS measurements and perform three 3 s measurements for each well.

6. Analyze the SNP genotypes using the software supplied with the SMFD system (examples of typing results are shown in *Fig. 2*).

Notes
[a]We perform both the first PCR and SSP–PCR in a thermal cycler capable of holding 384-well plates, as this enables us to perform high-throughput SNP analysis. Thermal cyclers capable of holding individual 0.2 or 0.5 ml tubes or 96-well plates are likely to be suitable.

[b]Other instruments and software may also be used for sequence analysis.

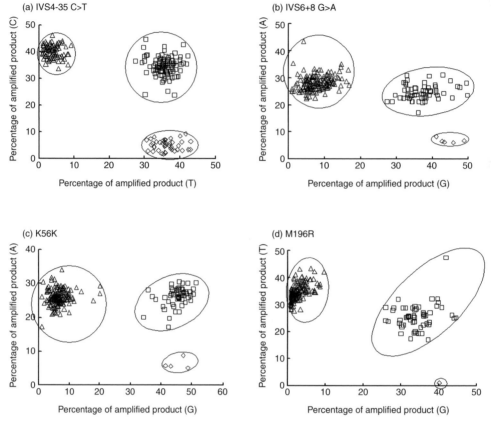

Figure 2. SNP typing results as determined by FCS using DOP–PCR products as templates.
A total of 216 samples was analyzed. The genomic DNA concentration was measured precisely using the PicoGreen method, and 10 ng of genomic DNA was used as the template for a 100 µl DOP–PCR. One microlitre of fivefold-diluted product was used as the template for a subsequent PCR. Genotypes were determined by sequencing of the genomic DNAs. (*a*) TNFR1 IVS4-35 C>T: ◇, TT; □, TC; △, CC. (*b*) TNFR1 IVS6+8 G>A: ◇, GG; □, GA; △, AA. (*c*) TNFR2 K56K: ◇, GG; □, GA; △, AA. (*d*) TNFR2 M196R: ◇, GG; □, GT; △, TT. The *x*-axis shows the percentage of amplified product for 633 nm (Cy5), while the *y*-axis shows the percentage of amplified product for 543 nm (TAMRA).

3.1. Results

3.1.1. Genome coverage

Cheung & Nelson (9) showed by microsatellite genotyping that a large proportion of the genome can be amplified by DOP–PCR. In their study, all 55 microsatellites tested were efficiently amplified from the DOP–PCR products. Our experience of SNP analysis also indicates that most regions in the genome can be amplified by DOP–PCR. We have succeeded in PCR amplification of 431 out of 441 SNPs (98%) using DOP–PCR products as PCR templates. Telenius *et al.* (1) demonstrated that a single degenerate primer can efficiently amplify DNA from the genomes of non-human species, including mouse and *Drosophila*.

3.1.2. Starting template DNA

Precise measurement and normalization of the amount of template DNA is important. DOP–PCR is performed in a 100 µl reaction mixture using 10 ng of genomic DNA as starting template. A shortage of genomic DNA template sometimes leads to a lower reliability of genotyping for some SNPs. Thus, the PicoGreen method is used for precise measurement of the genomic DNA concentration. The concentration of genomic DNA is adjusted with 10 mM Tris-HCl (pH 8.0) containing 0.1 mM EDTA.

3.1.3. DOP–PCR yield

DOP–PCR can amplify genomic DNA more than 100-fold. Using 10 ng of genomic DNA as a template for DOP–PCR, 500 different PCRs are possible from the resulting WGA product (see *Figs. 2* and *3*). SNP typing sometimes failed when we used lower amounts of DOP–PCR products, indicating that further dilution of DOP–PCR products results in reduced reliability of genotyping for some SNPs. Cheung & Nelson (9) showed that 40 ng of genomic DNA was amplified with DOP–PCR to an average of 8 µg (200-fold amplification) as determined by A_{260}. In the same study, all microsatellite markers tested were amplified 200–600-fold from 0.6–40 ng of genomic DNA.

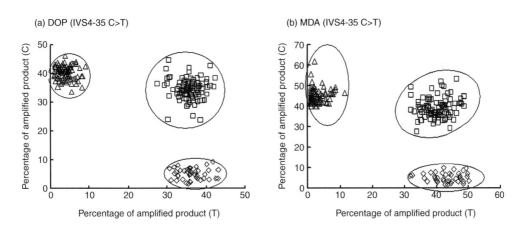

Figure 3. SNP typing results using DOP–PCR and MDA products as templates as determined by FCS (using precisely determined concentrations of genomic DNA).
SNP TNFR1 IVS4-35 C>T was analyzed in 216 samples. The genomic DNA concentration was measured precisely by the PicoGreen method. (*a*) Ten nanograms of genomic DNA was used as the template for a 100 µl DOP–PCR. One microlitre of fivefold-diluted product was used as the template for a subsequent PCR. (*b*) Ten nanograms of genomic DNA was used as the template for a 20 µl MDA reaction (using the GenomiPhi Kit, Amersham Biosciences, according to the manufacturer's instructions). One microlitre of 25-fold-diluted product was used as the template for a subsequent PCR. Genotypes were determined by sequencing of genomic DNAs. Samples are denoted by the corresponding genotype symbols as follows: ◇, TT; □, TC; △, CC. The *x*-axis shows the percentage of amplified product for 633 nm (Cy5), while the *y*-axis shows the percentage of amplified product for 543 nm (TAMRA).

3.1.4. DOP–PCR product size

The DOP–PCR products range from 200 to 1000 bp based on ethidium bromide staining of agarose gels (9). In our SNP typing studies, PCR primers were designed to produce amplification fragments up to 500 bp. Successful PCR amplifications using DOP–PCR products as template indicate that fragments of more than 500 bp in length can be obtained by DOP–PCR.

3.1.5. Amplification bias

Occasionally, we observe biased amplification of some heterozygous samples in mass SNP typing (see *Fig. 4a*, samples situated between clusters AA and CA, denoted by crosses). In this case, the amount of genomic DNA for some of the samples may not have been sufficient (see section 3.1.2). In addition, in microsatellite analysis, Grant *et al.* (10) noticed some preferential amplification of shorter alleles, although other reports have described equal amplification for microsatellites (9). It is important to take these points into consideration when using DOP–PCR-amplified DNA.

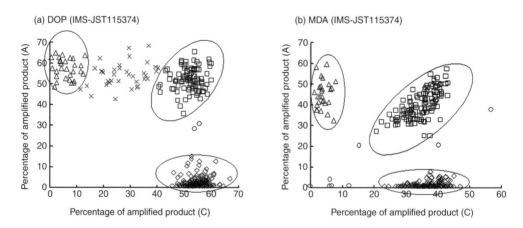

Figure 4. SNP typing results using DOP–PCR and MDA products as template as determined by FCS (using roughly measured concentrations of genomic DNA).
SNP IMS-JST115374 was analyzed in 246 samples. The genomic DNA concentration was roughly measured using a spectrophotometer. (*a*) Ten nanograms of genomic DNA was used as template for a 100 µl DOP–PCR. One microlitre of fivefold-diluted product was used as template for the subsequent PCR. (*b*) Ten nanograms of genomic DNA was used as template for a 20 µl MDA reaction (using the GenomiPhi Kit, Amersham Biosciences, according to the manufacturer's instructions). One microlitre of the 25-fold-diluted product was used as the template for the subsequent PCR. Samples are denoted by the corresponding genotype symbols as follows: ◇, CC; □, CA; △, AA; ○ and ×, not possible to judge. The *x*-axis shows the percentage of amplified product for 543 nm (TAMRA), while the *y*-axis shows the percentage of amplified product for 633 nm (Cy5).

3.1.6. Multiple-displacement amplification (MDA) provides greater accuracy in downstream genotyping assays

Recently, commercial kits (GenomiPhi, Amersham Biosciences; REPLI-g, Qiagen) employing MDA (see *Chapters 8–11*) have been used for WGA. When the amount of genomic DNA is sufficient (10 ng) for all samples, SNP typing using DOP–PCR products as template provides accurate results (see *Fig. 2*), comparable to those obtained using MDA products (see *Fig. 3*). However, we occasionally observed cases where MDA-generated DNA gave greater genotyping accuracy in mass SNP typing than DOP–PCR-generated DNA (see *Fig. 4*). In such cases, it is possible that suboptimal amounts of DNA are present in the DOP–PCR-amplified sample. When we compared DOP–PCR and MDA-amplified DNA for SNP typing, we succeeded in typing 82.1% (348 of 424) and 95.8% (68 out of 71) of the SNPs, respectively, when using the same genomic DNA template for WGA. For example, in *Fig. 4*, 34 out of 36 samples situated between clusters AA and CA (indicated by crosses) by DOP–PCR (see *Fig. 4a*) were classified in the CA cluster by MDA (see *Fig. 4b*).

3.2. Conclusion

SNP typing can be successfully performed using DOP–PCR-amplified DNA. However, it is important to ensure that sufficient starting template is used in the DOP–PCR and that an appropriate amount of DOP–PCR product is used for any subsequent PCRs. The genotypes determined by SSP–PCR and FCS using DOP–PCR samples were 100% in agreement with those determined by direct sequencing of genomic samples. Under these conditions, for most if not all cases, there should be no or very little biased amplification by DOP–PCR.

4. REFERENCES

★★★ 1. Telenius H, Carter NP, Bebb CE, Nordenskjold M, Ponder BA & Tunnacliffe A (1992) *Genomics*, **13**, 718–725. – *First report of DOP–PCR.*

2. Dietmaier W, Hartmann A, Wallinger S, *et al.* (1999) *Am. J. Pathol.* **154**, 83–95.

3. Hirose Y, Aldape K, Takahashi M, Berger MS & Feuerstein BG (2001) *J. Mol. Diagn.* **3**, 62–67.

4. Huang Q, Schantz SP, Rao PH, Mo J, McCormick SA & Chaganti RS (2000) *Genes Chromosomes Cancer*, **28**, 395–403.

5. Kittler R, Stoneking M & Kayser M (2002) *Anal. Biochem.* **300**, 237–244.

6. Telenius H, Pelmear AH, Tunnacliffe A, *et al.* (1992) *Genes Chromosomes Cancer*, **4**, 257–263.

7. Umayahara K, Numa F, Suehiro Y, *et al.* (2002) *Genes Chromosomes Cancer*, **33**, 98–102.

8. Kallioniemi A, Kallioniemi OP, Sudar D, *et al.* (1992) *Science*, **258**, 818–821.

★ 9. Cheung VG & Nelson SF (1996) *Proc. Natl. Acad. Sci. U. S. A.* **93**, 14676–14679. – *The use of DOP–PCR-amplified DNA in SNP genotyping.*

10. Grant SF, Steinlicht S, Nentwich U, Kern R, Burwinkel B & Tolle R (2002) *Nucleic Acids Res.* **30**, e125.

11. Bannai M, Higuchi K, Akesaka T, *et al.* (2004) *Anal. Biochem.* **327**, 215–221.

12. Jordan B, Charest A, Dowd JF, *et al.* (2002) *Proc. Natl. Acad. Sci. U. S. A.* **99**, 2942–2947.

CHAPTER 3

Whole genome amplification by improved primer-extension pre-amplification PCR

Peter J. Wild and Wolfgang Dietmaier

Institute of Pathology, University of Regensburg, Franz-Josef-Strauss-Allee 11, 93053 Regensburg, Germany

1. INTRODUCTION

Several WGA methods have been developed, allowing multiple molecular analyses from a few or even single cells. Using these techniques, the entire genome can be amplified, generating microgram quantities of DNA, which can then be analyzed by multiple approaches.

PCR-based WGA can be performed using either nondegenerate or degenerate primers with the latter being easy to use, as well as cost- and time-effective for DNA amplification. There are two approaches. One is degenerate-oligonucleotide-primed PCR (DOP–PCR) (1) (see Chapter 2), which is often used as the first step for *in situ* hybridization studies with either flow-sorted or microdissected chromosomes and for comparative genomic hybridization. DOP–PCR primers have defined sequences at their 5′ and 3′ ends, with a random hexamer sequence located between these regions. PCR amplification is performed under low-stringency conditions for the first five to eight cycles, followed by >28 cycles with a more stringent annealing temperature. A second method of WGA, first described by Zhang *et al.* (2), is known as primer-extension pre-amplification PCR (PEP–PCR). In contrast to DOP–PCR, totally degenerate PCR primers of 15 nt are used in PEP–PCR. In each of 50 cycles, the template is first denatured at 92°C. Primers are then annealed at a low-stringency temperature (37°C), which is then gradually increased to 55°C and held for 4 min for polymerase extension. *Fig. 1* presents a graphical representation of WGA using PEP–PCR.

1.1. Method of choice

Comparison of different WGA methods has shown that the efficiency (i.e. the frequency of successful downstream amplifications after pre-amplification) of

Whole Genome Amplification: *Methods Express* (S. Hughes and R. Lasken, eds.)

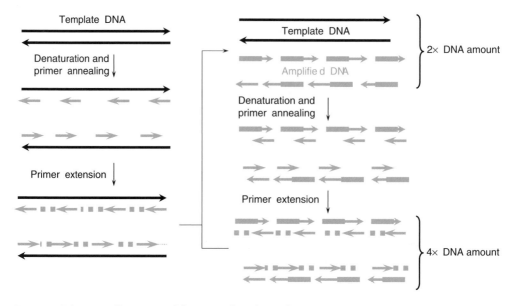

Figure 1. Schematic illustration of the principles of WGA by PEP–PCR.

PEP–PCR is higher than that of DOP–PCR (3, 4). Furthermore, the efficiency of PEP–PCR can be enhanced by several modifications, which is of particular interest when investigating tumor DNA from routine paraffin-embedded tissue sections. Improved PEP–PCR (I–PEP–PCR) is a reliable method for WGA, enabling multiple microsatellite (see *Fig. 2*) and sequencing studies with a few or even single cells. In combination with laser microdissection techniques, I–PEP–PCR provides a powerful tool to study the molecular changes underlying carcinogenesis, clonal expansion, and tumor dissemination.

I–PEP–PCR combines the methods of PEP–PCR and 'long PCR' (5) in which two different DNA polymerases are used in combination in the PCR. *Taq* DNA polymerase is used to carry out most of the primer extension as in a traditional PCR. A second proofreading DNA polymerase with 3'→5' exonuclease activity is used to remove misincorporated nucleotides that slow the progression of *Taq* DNA polymerase. When a nucleotide is misincorporated, such as a G opposite a T, the bases more readily melt into the single-stranded form, creating a non-productive 3' primer terminus. Because *Taq* DNA polymerase lacks its own 3'→5' exonuclease activity, it tends to stall after these misincorporations. Very low levels of the proofreading DNA polymerase are sufficient to remove the mismatched nucleotide and allow *Taq* DNA polymerase to proceed efficiently with primer extension. The result is a far more efficient PCR, the ability to amplify longer targets, and also increased fidelity due to the removal of the misincorporated nucleotides. Use of a long PCR system to carry out PEP–PCR (i.e. I–PEP–PCR)

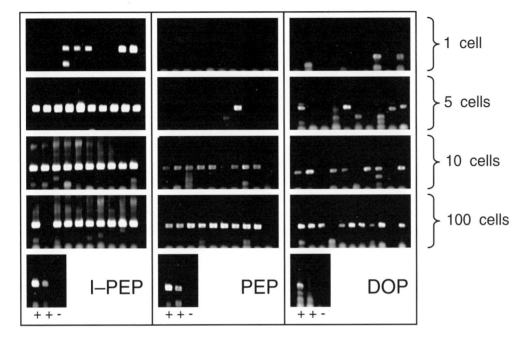

Figure 2. Results of downstream amplification reactions after I–PEP reflecting the efficiency of I–PEP–PCR compared with unmodified PEP–PCR and DOP–PCR.
A 1/30 aliquot of pre-amplified DNA was used for specific amplification of a 540 bp β-globin fragment. SW480 cells were sorted as single, five, ten, and 100 cells by fluorescence-activated cell sorting in ten parallel samples. Cells were lysed and the DNA was pre-amplified by I–PEP–PCR, PEP–PCR, or DOP–PCR. Positive controls (+) were 10 ng of genomic DNA either pre-amplified (left-hand lane) or not pre-amplified (right-hand lane). Negative controls (–) were H$_2$O instead of DNA in the pre-amplification PCR followed by specific PCR. Reprinted from (9) with permission of the *American Journal of Pathology*.

results in a dramatic improvement in the efficiency and fidelity of WGA (3, 6). It is important to note that I–PEP–PCR may not produce DNA products as long as those traditionally made in long PCR. Furthermore, I–PEP–PCR products apparently become shorter and shorter, with random priming to regions internal to amplicons (7). As with other PCR-based WGA methods (8), I–PEP–PCR generates mostly short DNA products (see *Fig. 3*) and can only be used for PCR genotyping assays that target segments of <400 bp.

This chapter will describe in detail the process of WGA by I–PEP–PCR as used in our laboratory. As an example of its utility, typical results obtained from fluorescence-based microsatellite and sequence analysis of microdissected specimens of colon cancer will be presented.

2. METHODS AND APPROACHES

2.1. Cell isolation

Molecular analysis in tumor pathology should be performed in precisely determined areas within homogeneous tumor cells. The analysis of tumor-specific genetic alterations can be compromised by the presence of normal cells. Thus, contamination by stromal and inflammatory cells should be minimized. For reliable microsatellite analysis and detection of chromosomal deletions by loss of heterozygosity (LOH) studies, a tumor cell content of at least 80% is required. In order to obtain such high cell content from small tissue samples, accurate tissue microdissection is important. The spectrum of microdissection techniques ranges from manual approaches (needle microdissection) to the use of a micro-manipulator and laser-assisted microdissection. These procedures are prerequisites for isolation of contamination-free, morphologically defined, pure cell populations from histologically heterogeneous tumor areas.

The DNA used in these experiments was extracted from histological tissue sections using laser microdissection. Investigation of laser-microdissected lesions, which provide only relatively low tumor cell numbers, requires pre-amplification of the DNA using WGA when multiple microsatellite markers are to be analyzed.

3. RECOMMENDED PROTOCOLS

Protocol 1

DNA isolation

Equipment and Reagents

- QIAamp DNA Mini Kit (Qiagen)
- 10× Expand HiFi PCR Buffer No. 3 (Roche)
- Proteinase K (20 mg/ml) (Merck)
- Tween 20 (Merck)
- Nuclease-free water (Sigma)
- Gelatine (PCR Optimization Kit; Roche)

Method

1. For isolation of DNA from 50–1000 cells[a], lyse the microdissected cells by the addition of 1 µl of 10× Expand HiFi Buffer No. 3 (final concentration 1×), 0.5% Tween 20 (final concentration) and 1 µl proteinase K (20 µg), made up to a final volume of 10 µl with nuclease-free water. Vortex the sample briefly and incubate for 4 h at 50°C, followed by 15 min at 94°C[b].

    or

    For isolation of DNA from more than 1000 cells, extract DNA using the QIAamp DNA Mini Kit, according to the manufacturer's specifications. Increase the DNA yield by eluting the DNA twice with 100 µl of 70°C pre-heated water[c,d]. Use a 10 µl aliquot (or 10–100 ng DNA) for the I–PEP–PCR.

2. Quantify the DNA yield using a standard spectrophotometer by measuring the absorbance at 260 nm.

Notes

[a]At least 200 cells should be prepared from each specimen to prevent preferential monoallelic amplification. In addition, to prevent false-positive results due to monoallelic amplification of nonspecific bands, all positive results should be validated.

[b]This lysate can be used directly for I–PEP–PCR.

[c]Each elution step includes a 5 min incubation of the QIAamp spin column with water pre-heated to 70°C before centrifugation at 16 000 g for 1 min.

[d]After elution of the DNA twice with 100 µl of water, the PCR template concentration can be further increased by reducing the elution volume to 50 µl using a SpeedVac (SC110; Savant) or by isopropanol precipitation in the presence of a glycogen carrier (final concentration 0.05 mg/ml).

Protocol 2

I–PEP WGA

Equipment and Reagents

- 280 µM 15mer random primer (5′-NNNNNNNNNNNNNNN-3′; Metabion)
- 25 mM $MgCl_2$
- Gelatine (1 mg/ml) (PCR Optimization Kit; Roche)
- 10× Expand HiFi PCR Buffer No. 3 (Roche)
- 10 mM dNTP mix
- Expand HiFi polymerase (3.6 units/µl; Roche)
- Nuclease-free water (Sigma)
- Thermocycler PTC-100 (MJ Research)
- SpeedVac SC110 (Savant)

Method

1. Prepare the I–PEP–PCR master mix by combining the following components for each reaction:

    3.4 µl of I–PEP primer (16 µM final concentration)
    6 µl of $MgCl_2$ (2.5 mM final concentration)
    3 µl of gelatine (0.05 mg/ml final concentration)
    6 µl of 10× Expand HiFi PCR Buffer No. 3
    0.6 µl of dNTP mix (0.1 mM final concentration)
    29.6 µl of nuclease-free water
    1.4 µl of Expand HiFi polymerase (0.084 units/µl final concentration)

2. Combine 50 µl of I–PEP–PCR master mix with 10 µl of DNA (or 10–100 ng of DNA) from *Protocol 1.*

3. Amplify the DNA in a thermocycler using either of the following PCR programs[a][b]:

    - Profile 1
      First denaturation step: 2 min at 94°C
      50 cycles: 1 min at 94°C
                               2 min at 28°C, then ramp at 0.1°C/s to 55°C
                               4 min at 55°C
                               30 s at 68°C
      Final elongation step: 8 min step at 68°C
    - Profile 2[c]
      First denaturation step: 4 min at 94°C
      20 cycles: 30 s at 94°C
                               1 min at 28°C, then ramp at 0.1°C/s to 55°C
                               45 s at 55°C
      30 cycles: 30 s at 94°C
                               45 s at 60°C
                               1 min at 72°C
      Final elongation step: 8 min step at 68°C

Notes

[a]For some downstream applications it may be necessary to both purify (using the Qiagen QIAquick PCR Purification Kit) and quantify (using a spectrophotometer or by PicoGreen quantification (Molecular Probes)) the amplified DNA before use.
[b]The amplification efficiency may be improved in some cases by adding dimethyl sulfoxide to a final concentration of 5% (Merck).

ᶜThe second I–PEP profile avoids the 4 min incubation step at 55°C and therefore reduces the time required for amplification. We have not noticed any appreciable difference in the DNA yield. The typical yield for I–PEP-PCR DNA is estimated to range from 50 to 1000 ng/60 μl, which is in accordance with other reports (10).

3.1. Downstream applications

The typical results obtained from an I-PEP-PCR reaction are shown in Figure 3.

3.1.1. Microsatellite analysis

At least two types of genetic instability can be found in cancer – microsatellite instability (MSI) and chromosomal instability. It has been shown that, in colorectal cancer, a set of five microsatellite markers should be analyzed to determine the high-frequency MSI phenotype (3). This primer set is recommended as an international standard primer panel for MSI analysis in colorectal cancer (11). The characteristics of these microsatellite primers are given in *Table 1*. MSI analysis can be performed using either standard (unlabeled) or fluorescently labeled primers. A multiplex microsatellite PCR kit, containing fluorescently labeled primers, is commercially available (HNPCC Microsatellite Instability Test; Roche) that permits the simultaneous testing of all five first-choice reference markers.

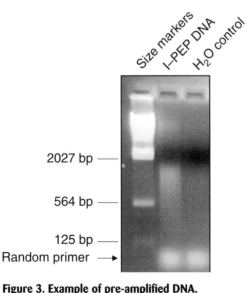

Figure 3. Example of pre-amplified DNA.
I–PEP DNA (3 μl) was separated by 1.3% agarose gel electrophoresis and visualized by ethidium bromide staining. Typically, a DNA smear ranging from 150 to 500 bp is generated. The maximum size of the amplified DNA can be up to 1.5 kb. Size markers were *Hind*III-digested λ DNA (Fermentas).

Table 1. Microsatellite primers used for MSI testing of colorectal cancers

Repeat type	Marker	Primer sequence (5′→3′)	Annealing temperature (°C)
Mononucleotide	BAT-25	F TCGCCTCCAAGAATGTAAGT R TCTGCATTTTAACTATGGCTC	58
	BAT-26	F TGACTACTTTTGACTTCAGCC R AACCATTCAACATTTTTAACCC	58
Dinucleotide	D5S346 (APC)	F ACTCACTCTAGTGATAAATCG R AGCAGATAAGACAGTATTACTAGTT	55
	D17S250 (Mfd15)	F GGAAGAATCAAATAGACAAT R GCTGGCCATATATATATTTAAACC	52
	D2S123	F AAACAGGATGCCTGCCTTTA R GGACTTTCCACCTATGGGAC	60

F, forward primer; R, reverse primer.

However, in some cases, separate amplification of microsatellites will be more robust and provide higher amplification rates. When using I–PEP-amplified DNA, we suggest using 2 µl of pre-amplified DNA (approximately 10–50 ng) for each microsatellite amplification. We strongly recommend performing control pre-amplification tests, using a polymorphic microsatellite PCR marker (e.g. D2S123), with different numbers of microdissected cells to determine the minimum amount of required cell material for each sample. This test will determine whether both alleles have been amplified during pre-amplification and help to exclude allele dropout effects. When using fluorescently labeled primers for detection of LOH and MSI, we use a capillary electrophoresis system (e.g. ABI PRISM 3100 Genetic Analyzer; Applied Biosystems). However, MSI can be detected using polyacrylamide gels if standard primers are used. MSI is defined by the presence of novel peaks after PCR amplification of tumor DNA that are not present in PCR products of the matching normal DNA (see *Fig. 4a*). LOH is defined as a decrease in signal intensity of the tumor sample allele to at least 50% relative to the matched normal DNA allele (see *Fig. 4b*).

3.1.2. DNA sequence analysis

Mutation analysis is widely used in cancer research. Sequence analysis of laser-microdissected tumor tissue is often problematic, since multiple exons of a given gene should be analyzed. For this reason, a protocol such as I–PEP-PCR allows complete mutation analysis of all exons of a gene, e.g. the tumor protein p53 gene (*TP53*), which is often found to be mutated in colorectal and many other cancers. Here, we have provided evidence for successful *TP53* sequence analysis. We used 2 µl of pre-amplified DNA (approximately 10–50 ng) for PCR amplification of exons 5, 6, 7, 8, and 9 from *TP53*. After polyethylene glycol precipitation, the PCR products were bidirectionally sequenced using BigDye Terminator chemistry (Applied Biosystems). An electropherogram generated by an ABI PRISM 3100 sequencer (Applied Biosystems) showing a heterozygous *TP53* mutation at codon 248 is shown in *Fig. 5*. The heterozygous base is denoted by N.

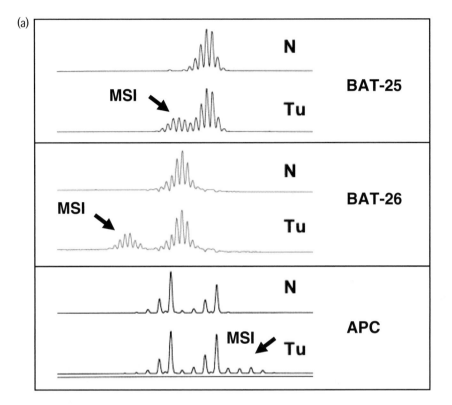

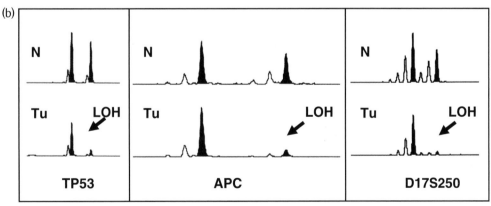

Figure 4. Examples of fluorescence-based microsatellite analysis by automatic fragment analysis (GENESCAN version 2.1) using ABI PRISM 3100 sequencing equipment.
Typical results obtained for MSI (*a*) and LOH (*b*) analysis of I–PEP–PCR-amplified matched normal and tumor DNA (arrows) are shown. The microsatellite markers used are indicated. N, DNA from normal colon mucosa; Tu, tumor DNA.

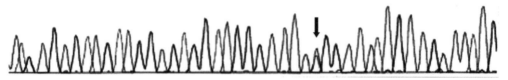

CAG T G T G AT G AT G G T G A G G AT G G G C C T C N G G T T CA T G C C G C C CAT G

Figure 5. p53 DNA sequence analysis.
A heterozygous mutation in codon 248 (C>T, Arg>Gln) within exon 7 is marked by an arrow.

4. TROUBLESHOOTING

- Successful I–PEP depends on high-quality starting DNA. If there is ineffective PCR amplification of a specific control gene with unamplified DNA, negative results after I–PEP are to be expected. For the initial evaluation of DNA quality, a standard control PCR (e.g. with D2S123 marker) and DNA from tissue that has not been laser microdissected should be performed prior to I–PEP.

- If I–PEP–PCR performance is poor, the amount of DNA for I–PEP–PCR should be increased stepwise to prevent inhibition of amplification by overloading effects.

- In some cases, it is necessary to try different amounts of pre-amplified DNA in downstream PCR amplification procedures. In our experience, the best downstream PCR efficiency can also be achieved by lowering the amount of I–PEP DNA in further PCR setups. We recommend using a series of four different DNA amounts, e.g. 0.5 (2.5–12.5 ng), 1 (5–25 ng), 2 (10–50 ng), and 4 µl (20–100 ng), in problematical downstream PCR amplifications.

5. REFERENCES

1. Telenius H, Pelmear AH, Tunnacliffe A, *et al.* (1992) *Genes Chromosomes Cancer*, **4**, 257–263.
★★★ 2. Zhang L, Cui X, Schmitt K, Hubert R, Navidi W & Arnheim N (1992) *Proc. Natl. Acad. Sci. U. S. A.* **89**, 5847–5851. – *First report of whole genome amplification by PEP–PCR.*
3. Dietmaier W, Wallinger S, Bocker T, Kullmann F, Fishel R & Ruschoff J (1997) *Cancer Res.* **57**, 4749–4756.
4. Sun G, Kaushal R, Pal P, *et al.* (2005) *Legal Medicine*, in press.
5. Barnes WM (1994) *Proc. Natl. Acad. Sci. U. S. A.* **91**, 2216–2220.
6. Zheng S, Ma X, Buffler PA, Smith MT & Wiencke JK (2001) *Cancer Epidemiol. Biomarkers Prev.* **10**, 697–700.
7. Buchanan AV, Risch GM, Robichaux M, Sherry ST, Batzer MA & Weiss KM (2000) *Hum. Biol.* **72**, 911–925.
8. Kittler R, Stoneking M & Kayser M (2002) *Anal. Biochem.* **300**, 237–244.
9. Dietmaier W, Hartmann A, Wallinger S, *et al.* (1999) *Am J. Pathol.* **154**, 83–95.
10. Coombes MM, Mao L, Steck KD, Luna MA & El-Naggar AK (1998) *Diagn. Mol. Pathol.* **7**, 197–201.
11. Boland CR, Thibodeau SN, Hamilton SR, *et al.* (1998) *Cancer Res.* **58**, 5248–5257.

CHAPTER 4

Global amplification using SCOMP: single-cell comparative genomic hybridization

Nona C.R. Arneson[1], Arezou A. Ghazani[2] and Susan J. Done[1,2]

[1]Ontario Cancer Institute, Division of Applied Molecular Oncology, University Health Network, Toronto, Ontario, Canada; [2]Department of Laboratory Medicine and Pathobiology, University of Toronto, Toronto, Ontario, Canada

1. INTRODUCTION

WGA techniques have recently come to the forefront of modern molecular research due to the advent of high-throughput microarray technology. Microarray-based comparative genomic hybridization (CGH) methods are now being widely used for high-resolution genomic analysis (1). Several platforms for CGH microarray analysis have been developed including cDNA arrays (2), oligonucleotide arrays (3), bacterial artificial chromosome (BAC) arrays (4–6), and, most recently, single nucleotide polymorphism (SNP) arrays, all of which have been used to identify genomic amplifications and deletions (7, 8). Unfortunately, accurate and reproducible array CGH requires an ample source of high-quality genomic DNA. Using clinical samples often means obtaining DNA by microdissection from small amounts of fixed archival tissues, which results in a low quantity and poor-quality genomic template for analysis. PCR-based WGA methods have been validated and successfully used for both high- and low-resolution genomic analysis. One example is degenerate-oligonucleotide-primed PCR (DOP–PCR), which has been used for chromosomal CGH (9–15), SNP genotyping (16, 17), microsatellite genotyping (18–20), mutation detection (17), and even array CGH (21, 22). However, the reproducibility and fidelity of DOP–PCR has often been brought into question. Recently developed WGA methods such as those based on multiple displacement amplification (MDA) using φ29 DNA polymerase (23, 24) have shown excellent reproducibility and fidelity when tested with high-resolution methods such as array CGH (25, 26). However, this technique may not be suitable for use with formalin-fixed, paraffin-embedded (FFPE) clinical

Whole Genome Amplification: *Methods Express* (S. Hughes and R. Lasken, eds.)

samples due to the fragmentation of the template genomic DNA (24). This chapter outlines another method of global amplification, based on the concept of single-cell comparative genomic hybridization (SCOMP), which uses genomic representations (3) to study the genomic DNA of a single cell. This method was originally designed for chromosomal CGH of single cells, but has the potential to be used for a variety of molecular analyses, including array CGH, and has been successfully applied for use with moderately degraded archival specimens.

2. METHODS AND APPROACHES

2.1. SCOMP

SCOMP was originally published by Klein *et al.* (27) for use in the genetic analysis of single cells. The significance of the technique was highlighted in this publication by the finding of different (yet mostly similar) CGH profiles when this method was performed on several single cells isolated from a culture of MCF7 breast cancer cells. CGH analysis using pooled DNA (bulk extraction) failed to reveal the additional aberrations shown by some individual cells. These aberrations were confirmed by fluorescent *in situ* hybridization providing evidence that they were not PCR artifacts. In most laboratories, even when microdissection is used, DNA is still obtained and subsequently analyzed using bulk extraction. If rare genotypes are responsible for the evolution of disease, then they would be overlooked unless analyzed at the single-cell level. Klein *et al.* aimed to develop a method whereby the genome of a single cell could be reliably amplified to provide enough material for a comprehensive analysis of the genome. Using CGH, loss of heterozygosity (LOH), and mutation analysis, Klein *et al.* confirmed the fidelity of this strategy and have since used it to study single cells isolated from bone marrow (28), DNA extracted from archival material (29), single disseminated tumor cells in minimal residual cancer (30), and circulating melanoma cells (31). Although SCOMP has been validated using chromosomal CGH, it is desirable to use techniques such as array CGH, which have a higher resolution and better dynamic range, to analyze the genomic signatures of small microdissected tissues. In this chapter, additional protocols are described using the SCOMP amplification procedure in conjunction with array CGH for high-resolution analysis of genomic DNA obtained from high-quality sources, as well as from moderately degraded fixed tissues.

2.2. Principles of SCOMP

SCOMP begins by converting the genome to a high-complexity resolution (3) with a fragment size of less than 2 kb by restriction enzyme digestion. Several restriction enzymes were tested in developing this protocol; however, only *Mse*I was able to produce a smear in the expected range of 100–1500 bp (27). Following enzyme digestion, adaptors containing primer sequences are ligated on to the ends of the genomic DNA and subsequently amplified in a ligation-mediated PCR (32, 33). The reaction volume is purposely kept to a minimum and all buffers are optimized to eliminate the need to purify the reaction between steps. As a result, enzymatic digestion, adaptor ligation, and PCR are all completed in a single tube, resulting in minimal manipulation and negligible sample loss.

The PCR products resulting from SCOMP have been used for chromosomal CGH, array CGH, LOH analysis, and direct sequencing. Chromosomal CGH analysis requires the SCOMP products to be labeled in an additional PCR with appropriate dyes. For array CGH analysis, the SCOMP products are directly labeled by random-primer labeling using either Cy3 or Cy5 fluorescent dye. Regardless of the downstream assay, it is essential to treat both the DNA of interest and the control genomic DNA in the same manner. SCOMP reduces the genome to only a representation, presumably because some of the sequences may fail to amplify due to large (or very small) amplicon size or because gene-specific target sequences may have the MseI restriction endonuclease site within them. Moreover, it is also important that the control DNA is of similar quality to the DNA of interest. For example, if the DNA of interest was extracted from FFPE material, then the control DNA should also be from this source, and preferably from the same specimen.

2.3. DNA template requirements

1. A single cell can be used to obtain genomic DNA for global amplification for some applications.
2. As little as 1 ng of genomic DNA template can be used for successful global amplification for array CGH. (Note that the SCOMP protocol for WGA for array CGH has not been tested using single cells.)
3. Degraded templates must be used with caution. Successful amplification will occur but may result in uneven coverage of the genome.
4. The amount of DNA synthesized in this method of global amplification is independent of the amount of DNA template added. In general, each reaction results in 2–3 µg of amplified DNA.

3. RECOMMENDED PROTOCOLS

Genomic DNA may be extracted by several methods. For microdissection of tissues embedded in paraffin, the sections must be deparaffinized prior to microdissection. For single-cell isolations and for microdissected material, it is recommended that sample handling be minimized as much as possible by using *Protocol 1*, which does not require further purification of the genomic DNA. Bulk samples of cells and tissues (including large areas or multiple sections acquired by microdissection) can be purified using a variety of commercially available methods such as the QIAamp DNA Mini Kit (Qiagen), as in *Protocol 2*. A flow diagram of the methods involved is shown in *Fig. 1*.

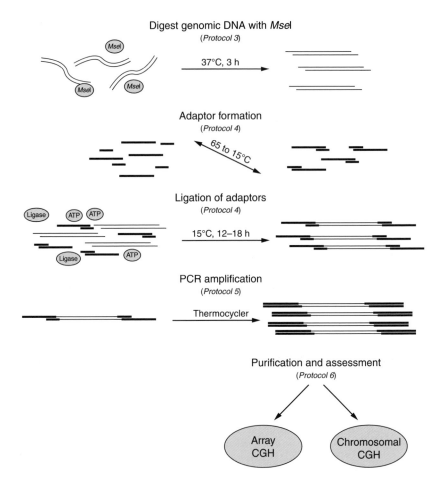

Figure 1. Schematic of SCOMP.

SCOMP is performed in several steps using a single tube to minimize sample handling and avoid contamination. The genomic DNA is digested with *Mse*I (*Protocol 3*) and adaptor complexes are subsequently ligated to the fragment ends (*Protocol 4*). Sequences in the adaptors are used for specific PCR amplification (*Protocol 5*). Products are purified and assessed by agarose gel electrophoresis and used in downstream applications (*Protocol 6*).

Protocol 1

DNA extraction from single cells and minute microdissected tissue (27, 34)

Equipment and Reagents

- Phosphate-buffered saline (PBS)
- One-Phor-All Buffer Plus (Amersham Biosciences)
- Tween 20 (Sigma)
- Igepal/Nonidet-P40 (Sigma)
- Proteinase K (10 mg/ml)
- Sterile, purified water (Sigma)
- Microcentrifuge
- Block heater or thermocycler
- Single-cell isolation buffer (1× PBS; 0.5% Igepal)
- Proteinase K digestion buffer (0.30 µl of One-Phor-All Buffer Plus; 0.13 µl of 10% Tween 20; 0.13 µl of 10% Igepal; 0.26 µl of proteinase K (10 mg/ml); 2.18 µl of H_2O; per sample)

Method

1. Transfer single whole cells, obtained by micromanipulation, into 1 µl of single-cell isolation buffer in a PCR tube.

2. Add 2 µl of proteinase K digestion buffer and incubate the reaction for 15 h at 42°C[a].

3. Heat inactivate the proteinase K by placing the reaction at 80°C for 10 min.

4. Spin the reaction in a microcentrifuge to collect droplets formed by evaporation.

Notes

[a]Tissues acquired by laser-capture microdissection should be placed directly in contact with 3 µl of proteinase K digestion buffer and incubated for 15 h at 56°C.

Protocols 3–7 were optimized using DNA from cell lines (commercially available from the American Type Culture Collection (ATCC)) and serial sections of histological specimens that were microdissected using an 18-gauge needle and purified using the QIAamp DNA Mini Kit.

Protocol 2

DNA extraction from bulk tissues

Equipment and Reagents
■ QIAamp DNA Mini Kit (Qiagen)
■ Sterile, purified water (Sigma)
■ 1% agarose gel containing 10 ng/ml ethidium bromide
■ 6× loading dye (Ficoll-400 (Amersham Biosciences); 0.25% Orange G (Sigma))
■ Equipment and reagents for agarose gel electrophoresis including 1× TBE agarose gel running buffer (89 mM Tris-HCl (pH 8.5); 89 mM boric acid; 2 mM EDTA)
■ UV light source and gel documentation system
■ Spectrophotometer
■ Microcentrifuge
■ Block heater or thermocycler

Method
1. Extract genomic DNA from tissue sections using the QIAamp DNA Mini Kit following the manufacturer's instructions[a] with a modification to the elution step.

2. Elute the genomic DNA in purified water to prevent interference in subsequent reaction steps due to buffer incompatibilities[b].

3. Quantify genomic DNA[c] by spectrophotometry or another method.

4. Analyze 200–300 ng by agarose gel electrophoresis to assess the DNA quality and degree of degradation[d].

Notes

[a]When digesting tissues that have been embedded in paraffin, it is beneficial to extend the digestion time from 15 up to 72 h with the addition of extra proteinase K after 12 h if tissue fragments are still visible in the tube. Although thorough mixing is required, vortexing of tissues can cause more damage to already degraded DNA and is not recommended.

[b]The pH of the water should be checked as recommended by Qiagen to ensure complete elution of the purified DNA. Decreasing the elution volume to 30–50 μl when using the Qiagen DNA Mini Kit is recommended to prevent unnecessarily diluting the genomic DNA when small quantities are expected. A second elution can increase DNA yield, but it is recommended to keep the two elutions in separate tubes until they are quantified.

[c]A NanoDrop ND-1000 UV-VIS Spectrophotometer (NanoDrop Technologies) is recommended for DNA quantitation as it requires only 1 μl of undiluted material for accurate measurement.

[d]The quality of the genomic DNA extracted from fixed tissues will have a direct effect on the results. The more degraded the sample, the less reliable it will be for analysis with this method. Not all tissues will be suitable.

Protocol 3

Stage I of the SCOMP protocol – *Mse*I digest

Equipment and Reagents
- 0.2 ml PCR microtube
- One-Phor-All Buffer Plus (Amersham Biosciences)
- *Mse*I restriction endonuclease (20 units/µl; New England Biolabs)
- Sterile, purified water (Sigma)
- Heat block or thermocycler
- Microcentrifuge

Method
1. Combine the following reagents in a 0.2 ml PCR tube (total reaction volume 5 µl), mix well by pipetting up and down, and incubate at 37°C for 3 h[a]:
     0.2 µl of One-Phor-All Buffer Plus
     0.2 µl of *Mse*I
     3 µl of DNA template[b]
     1.6 µl of water

2. Inactivate the *Mse*I by incubating the reaction at 65°C for 5 min.

3. Spin the reaction in a microcentrifuge to collect droplets formed by evaporation.

Notes
[a]If several SCOMP reactions are to be generated for subsequent analysis (i.e. more than 2–3 µg is required), it is possible to multiplex this step by combining several digests in a single tube.

[b]The 3 µl reaction from the proteinase K digest, from either single-cell isolations or laser-capture microdissected tissues (see *Protocol 1*), can be added directly after proteinase K inactivation without any purification. If using bulk-extracted genomic DNA (see *Protocol 2*), the genomic DNA concentration should be in the range of 10–100 ng/µl. A negative control using 3 µl of purified water should also be used to monitor for contamination. If desired, a positive control using 10 ng of good-quality genomic DNA may also be used to monitor successful PCR.

Following the *Mse*I digest of the genomic DNA, the adaptor is ligated to the resulting 5′ overhang. The complementary adaptor sequence is used as a primer acceptor site in the subsequent PCR. Concentrated enzymes are required for all of the following stages.

Protocol 7 describes DNA labeling of SCOMP products generated from bulk-extracted genomic DNA for chromosomal CGH. The methods for labeling and slide treatment are as described by Stoecklein *et al.* (29) and Zielenska *et al.* (35) respectively, with modifications to optimize labeling and detection of samples.

Protocol 4

Stage 2 of the SCOMP protocol – adaptor ligation

Equipment and Reagents
- One-Phor-All Buffer Plus (Amersham Biosciences)
- 100 µM LIB1 oligonucleotide (5′-AGTGGGATTCCTGCTGTCAGT-3′)
- 100 µM ddMseI1 oligonucleotide (5′-TAACTGACAGCdd-3′)
- 10 mM ATP (Roche)
- T4 DNA ligase (5 units/µl; Roche)
- Sterile, purified water (Sigma)
- Thermocycler

Method

1. Combine the following in a 0.2 ml PCR tube (total reaction volume 3 µl) for each sample in *Protocol 3*, including the negative control:

   0.5 µl of One-Phor-All Buffer Plus
   0.5 µl of LIB1 oligonucleotide
   0.5 µl of ddMseI1 oligonucleotide
   1.5 µl of water

2. Incubate the sample using a programmed temperature gradient in a thermocycler from 65 to 15°C ramping at 1°C/min to form the adaptor complexes.

3. Leaving the samples at 15°C, add 1 µl of ATP and 1 µl of T4 DNA ligase to each reaction.

4. Add the total volume (5 µl) of the *MseI*-digested genomic DNA (see *Protocol 3*, step 1), mix well by pipetting up and down, and incubate overnight (~12–16 h) at 15°C.

5. Spin reactions in a microcentrifuge to collect droplets formed by evaporation.

3.1. Downstream applications

3.1.1. Chromosomal CGH

Chromosomal CGH is carried out using standard procedures (29, 35). The only modifications to standard protocols are the use of 8 µg of the biotin-16-dUTP-labeled test and 8 µg of the digoxigenin-11-dUTP-labeled reference DNA for hybridization, as well as treatment of the metaphase chromosome slides with 10% pepsin/0.01 M HCl for 5 min instead of with proteinase K (commonly used in chromosomal CGH methods). Proteinase K is much more robust than 10% pepsin and easily causes overdigestion of the tissue resulting in poor hybridization. An additional washing step after hybridization is recommended due to the increased background that may be generated when using smaller probe sizes for chromosomal CGH. We recommend first washing the slides three times in 0.1% sodium dodecyl sulfate, 0.1× saline sodium citrate (pre-warmed to 60°C) for 5 min each with shaking before carrying out the standard washing procedures. A typical result obtained using SCOMP-amplified DNA for metaphase CGH, focusing on chromosome 17, is shown in *Fig. 3a* (in color section).

Protocol 5

Stage 3 of the SCOMP protocol – PCR amplification

Equipment and Reagents
- Expand Long Template PCR System[a] (Roche)
- 10 mM dNTP mix (2.5 mM each of dATP, dCTP, dGTP, and dTTP) (Invitrogen)
- Sterile, purified water (Sigma)
- Thermocycler

Method
1. Prepare a PCR master mix with the following reagents for each reaction in *Protocol 4*, including the negative control:

   3 µl of Expand Long Template Buffer 1
   2 µl of 10 mM dNTPs
   1 µl of Expand Long Template PolMix (3.5 units/µl)
   34 µl of water

2. Add 40 µl of the PCR master mix to the ligated genomic DNA fragments and place in a thermocycler with the following program[b]:

Number of cycles	Program
1	3 min[c] for 68°C
15	40 s at 94°C; 30 s at 57°C; 1 min 30 s + 1 s/cycle at 68°C
8	40 s at 94°C; 30 s at 57°C + 1°C/cycle ; 1 min 45 s + 1 s/cycle at 68°C
22	40 s at 94°C; 30 s at 65°C; 1 min 53 s + 1 s/cycle at 68°C
1	3 min 40 s at 68°C
1	Hold at 4–8°C

Notes
[a]Other PCR systems designed for high-fidelity PCR may be used in this protocol but have not been tested.
[b]The thermocycler program was provided by Dr. C. Klein. Further optimization of these cycling parameters has not been tested.
[c]The initial incubation at 68°C is required to fill in the recessive 3′ end of the lower DNA strand to generate a complementary primer sequence.

3.1.2. Array CGH

The purified PCR fragments can be used as a substitute for genomic DNA for array CGH. The amount labeled will depend on the labeling method and array platform used and should be optimized. As previously stated, the control genomic DNA should be prepared in the same manner, SCOMP-amplified, and be of similar quality to that of the test DNA. If desired, the adaptor sequences may be removed prior to cleaning the PCR products by *Tru*I restriction endonuclease digestion. *Tru*I is an isoschizomer of *Mse*I, but cuts efficiently in the PCR buffer, whereas *Mse*I

Protocol 6

Assessment, cleaning and quantification of PCR products

Equipment and Reagents
- 1% Agarose gel containing 10 ng/ml ethidium bromide
- 6× Loading dye (Ficoll-400 (Amersham Biosciences); 0.25% Orange G (Sigma))
- Equipment and reagents for agarose gel electrophoresis including 1× TBE agarose gel running buffer (89 mM Tris-HCl (pH 8.5); 89 mM boric acid; 2 mM EDTA)
- QIAquick PCR Purification Kit (Qiagen)
- Spectrophotometer

Method
1. Run 5–10 µl of the PCR products including the negative control from *Protocol 5* on a 1% agarose gel to assess fragment size and success of the reaction[a] (see *Fig. 2*).

2. Clean each sample using the QIAquick PCR Purification Kit according to the manufacturer's instructions with a modification to the elution step.

3. Elute the PCR products in 50 µl water[b].

4. Quantify the products by spectrophotometry.

Notes

[a]No amplification product should be obtained in the negative control lane. Ideally, fragments generated from a good-quality genomic DNA sample should range from 100 to 1500 bp. The more degraded the original genomic DNA sample, the smaller the PCR fragments will be.

[b]It is recommended that samples be eluted in water. If higher concentrations are required for downstream applications, the elution volume can be decreased from 50 to 30 µl, as recommended by Qiagen.

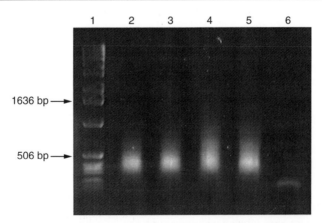

Figure 2. Agarose gel electrophoresis of SCOMP-amplified genomic DNA.
Typical results expected from the SCOMP procedure using different qualities of starting genomic DNA. (*Lane 1*) 1 kb DNA ladder (Invitrogen); (*lanes 2 and 3*) SCOMP products from DNA extracted from microdissected fixed material; (*lane 4*) SCOMP products from 10 ng of genomic DNA extracted from the cell line UACC-812 (ATCC CRL-1897); (*lane 5*) SCOMP products from 10 ng of DNA extracted from human placenta (Sigma); (*lane 6*) negative control (water).

Protocol 7

DNA labeling for chromosomal CGH using SCOMP-amplified DNA

Equipment and Reagents

- 1 nmol/µl Biotin-16-2'-deoxyuridine-5'-triphosphate (biotin-16-dUTP) (Roche)
- 1 nmol/µl Digoxigenin-11-2'-deoxyuridine-5'-triphosphate (digoxigenin-11-dUTP) (Roche)
- 10 µM LIB1 oligonucleotide (5'-AGTGGGATTCCTGCTGTCAGT-3')
- dNTP mix: dGTP, dATP, dCTP (10 mM each); dTTP (8.75 mM) (Invitrogen)
- Expand Long Template PCR System (Roche)
- *Tru*I restriction endonuclease (50 units/µl; Fermentas)
- Sterile, purified water (Sigma)

Method

1. Combine the following reagents in a 0.2 ml PCR tube:
   300–400 ng of the amplified DNA (in 2 µl) (see *Protocol 6*)[a]
   4 µl of Expand Long Template Buffer 1
   6 µl of LIB1 oligonucleotide
   1.4 µl of dNTP mix
   1.75 µl of digoxigenin-11-dUTP for reference DNA, *or*
   1.75 µl of biotin-16-dUTP for test DNA
   2 µl of Expand Long Template PolMix (3.5 units/µl)
   25 µl of water

2. Incubate the samples using the following PCR program:

Number of cycles	Program
1	1 min at 94°C, 30 s at 60°C, 2 min at 72°C
10	30 s at 94°C, 30 s at 60°C, 2 min + 20 s/cycle at 72°C
1	Hold at 8°C

3. Add 2 µl of *Tru*I enzyme and 2 µl of Buffer R+ to each PCR to remove LIB1 primer sequences.

4. Incubate at 65°C for 3 h.

5. Purify the PCR products using the QIAquick PCR Purification Kit according to the manufacturer's instructions.

Notes

[a]We have found that, depending on the quality of the starting DNA material used in the SCOMP protocol, 2 µl of product is approximately 300–400 ng of DNA. If good-quality DNA is used (i.e. extracted from a cell line or frozen tissue), 1 µl of the purified SCOMP-amplified PCR product may be sufficient. However, we suggest when first using this protocol that it is important to quantify the DNA yield, as further optimization may be required.

does not. However, *Tru*I cannot be substituted in the initial enzyme digest because it cannot be heat inactivated. As a greater amount of DNA may be required for multiple replicate experiments, it is recommended that several SCOMP reactions be pooled together to ensure that enough DNA is available. This may also reduce the effects of any resulting variability or PCR artifacts that may occur during the PCR amplification step. A comparison of the CGH profiles obtained when using SCOMP-amplified or non-amplified genomic DNA for the array showed no distinguishable differences and the regions of amplification were clearly detected using both templates (see *Fig. 3b* in color section).

Optimization of the quantity of amplified DNA required for probe labeling, good hybridization, and a strong signal are dependent upon the chosen array platform. We have used a cDNA array platform for array CGH and the methods have been tested on both good-quality genomic DNA and DNA obtained from FFPE tissues. It is important to note that the quality of the DNA from fixed tissues has a significant effect on the success of array CGH, regardless of whether SCOMP is applied. For this reason it is important to determine the tolerance of both the WGA method and the array platform when combining these for analyzing DNA from fixed tissues. Use of poor-quality genomic DNA may be overcome by using more genomic DNA template in the SCOMP protocol or by labeling more of the SCOMP products for array CGH; however, for more heavily degraded DNA, this may not be sufficient and must be tested individually.

4. TROUBLESHOOTING

- It is important to include a negative control in all steps of the SCOMP procedure to ensure that none of the reaction constituents is contaminated. If amplification products are observed in the negative control lane, the procedure should be repeated with new reagents.
- A positive control (such as 50 ng of good-quality genomic DNA) can also be used to ensure that the reaction is working optimally.
- It is recommended that dNTPs and oligonucleotides are stored as small aliquots, as repeated freeze–thaw cycles can affect the integrity of these reagents and thereby affect the efficiency of SCOMP.
- Best results are obtained when starting with good-quality genomic DNA; however, the procedure will successfully amplify lower-quality genomic DNA such as that obtained from fixed tissues. Keep in mind, however, that this procedure does not repair or improve the quality of the genomic DNA.
- The starting amount of genomic DNA is less critical than the quality; however, if the sample is known to be of low quality, increasing the amount of starting material is recommended.
- The size of the amplification product is directly dependent on the quality of the starting genomic DNA. The more fragmented the starting material, the smaller the amplification product smear will be. Ideally, it should be in the range of 100–1500 bp. The size range of the amplification products will have an effect on downstream experiments.

5. REFERENCES

1. Albertson DG & Pinkel D (2003) *Hum. Mol. Genet.* **12** (Review Issue 2), R145–152.
2. Pollack JR, Perou CM, Alizadeh AA, *et al.* (1999) *Nat. Genet.* **23**, 41–46.
★★ 3. Lucito R, Nakimura M, West JA, *et al.* (1998) *Proc. Natl. Acad. Sci. U. S. A.* **95**, 4487–4492. – *The original publication describing the use of genomic representations for molecular analysis.*
4. Pinkel D, Segraves R, Sudar D, *et al.* (1998) *Nat. Genet.* **20**, 207–211.
5. Snijders AM, Nowak N, Segraves R, *et al.* (2001) *Nat. Genet.* **29**, 263–264.
6. Ishkanian AS, Malloff CA, Watson SK, *et al.* (2004) *Nat. Genet.* **36**, 299–303.
7. Bignell GR, Huang J, Greshock J, *et al.* (2004) *Genome Res.* **14**, 287–295.
8. Zhao X, Li C, Paez JG, *et al.* (2004) *Cancer Res.* **64**, 3060–3071.
9. Ottesen AM, Skakkebaek NE, Lundsteen C, Leffers H, Larsen J & Rajpert-De Meyts E. (2003) *Genes Chromosomes Cancer,* **38**, 117–125.
10. Beheshti B, Vukovic B, Marrano P, Squire JA & Park PC (2002) *Cancer Genet. Cytogenet.* **137**, 15–22.
11. Harada T, Okita K, Shiraishi K, *et al.* (2002) *Oncology,* **62**, 251–258.
12. Hirose Y, Aldape K, Takahashi M, Berger MS & Feuerstein BG (2001) *J. Mol. Diagn.* **3**, 62–67.
13. Larsen J, Ottesen AM, Lundsteen C, Leffers H & Larsen JK (2001) *Cytometry,* **44**, 317–325.
14. Huang Q, Schantz SP, Rao PH, Mo J, McCormick SA & Chaganti RS (2000) *Genes Chromosomes Cancer,* **28**, 395–403.
15. Verhagen PC, Zhu XL, Rohr LR, *et al.* (2000) *Cancer Genet. Cytogenet.* **122**, 43–48.
16. Grant SF, Steinlicht S, Nentwich U, Kern R, Burwinkel B & Tolle R (2002) *Nucleic Acids Res.* **30**, e125.
17. Barbaux S, Poirier O & Cambien F (2001) *J. Mol. Med.* **79**, 329–332.
18. Kittler R, Stoneking M & Kayser M (2002) *Anal. Biochem.* **300**, 237–244.
19. Kim SH, Godfrey T & Jensen RH (1999) *J. Urol.* **162**, 1512–1518.
20. Cheung VG & Nelson SF (1996) *Proc. Natl. Acad. Sci. U. S. A.* **93**, 14676–14679.
21. Peng DF, Sugihara H, Mukaisho K, Tsubosa Y & Hattori T (2003) *J. Pathol.* **201**, 439–450.
22. Daigo Y, Chin SF, Gorringe KL, *et al.* (2001) *Am. J. Pathol.* **158**, 1623–1631.
23. Dean FB, Hosono S, Fang L, *et al.* (2002) *Proc. Natl. Acad. Sci. U. S. A.* **99**, 5261–5266.
24. Lage JM, Leamon JH, Pejovic T, *et al.* (2003) *Genome Res.* **13**, 294–307.
25. Wong KK, Tsang YT, Shen J, *et al.* (2004) *Nucleic Acids Res.* **32**, e69.
26. Wang G, Maher E, Brennan C, *et al.* (2004) *Genome Res.* **14**, 2357–2366. – *DNA amplification method tolerant to sample degradation.*
★ 27. Klein CA, Schmidt-Kittler O, Schardt JA, Pantel K, Speicher MR & Riethmuller G (1999) *Proc. Natl. Acad. Sci. U. S. A.* **96**, 4494–4499. – *The original publication first describing SCOMP.*
★★ 28. Klein CA, Seidl S, Petat-Dutter K, *et al.* (2002) *Nat. Biotechnol.* **20**, 387–392. – *SCOMP applied to clinical samples.*
★★ 29. Stoecklein NH, Erbersdobler A, Schmidt-Kittler O, *et al.* (2002) *Am. J. Pathol.* **161**, 43–51. Erratum **163**, 2645. – *Comparison showing that the fidelity of SCOMP is superior to DOP-PCR for CGH.*
★★ 30. Klein CA, Blankenstein TJ, Schmidt-Kittler O, *et al.* (2002) *Lancet* **360**, 683–689. – *SCOMP applied to clinical samples.*
31. Ulmer A, Schmidt-Kittler O, Fischer J, *et al.* (2004) *Clin. Cancer Res.* **10**, 531–537.
32. Mueller PR & Wold B (1989) *Science,* **246**, 780–786.
33. Pfeifer GP, Steigerwald SD, Mueller PR, Wold B & Riggs AD (1989) *Science,* **246**, 810–813.
34. Done SJ, Arneson NC, Ozcelik H, Redston M & Andrulis IL (1998) *Cancer Res.* **58**, 785–789.
★★ 35. Zielenska M, Bayani J, Pandita A, *et al.* (2001) *Cancer Genet. Cytogenet.* **130**, 14–21. – *The preferred protocol for CGH used in this chapter.*

CHAPTER 5

PRSG, a whole genome amplification method based on adaptor-ligation PCR of randomly sheared genomic DNA

Hiroki Sasaki and Kazuhiko Aoyagi

Genetics Division, National Cancer Center Research Institute, 1-1, Tsukiji 5-chome, Chuo-ku, Tokyo 104-0045, Japan

1. INTRODUCTION

To provide the quality and quantity of DNA required for large-scale genotyping studies, mutational analyses or copy number estimations, a renewable source of patient DNA is required. One approach for this is the establishment of Epstein–Barr virus-transformed B-cell lines from whole blood, but this is a time-consuming process and is thus not ideally suited for high-throughput analysis. In addition, the process of immortalization can introduce genetic alterations, which could possibly affect downstream applications. An alternative approach for the generation of microgram quantities of genome-representative DNA is the use of WGA. Since the early 1990s, several WGA methods have been reported, including primer-extension pre-amplification PCR (PEP–PCR), degenerate-oligonucleotide-primed PCR (DOP–PCR) (1–4) and multiple displacement amplification (MDA) (5). Although PEP–PCR and DOP–PCR are capable of generating sufficient quantities of DNA for mutational analysis and comparative genomic hybridization (CGH) from as few as 10–100 cultured or microdissected cells (6–8), biases constituting four to six orders of magnitude have been reported (5). The amplification bias for MDA-based WGA has been demonstrated to be only sixfold (5) when compared with the original genomic DNA. However, the requirement for high-molecular-weight DNA makes this approach unsuitable when using DNA extracted from fixed tissues, which tends to be of low molecular weight. As a result, a method of WGA is required that introduces little or no amplification bias, is suitable for use on DNA from fixed tissue, and can generate DNA for high-throughput genetic studies.

Whole Genome Amplification: *Methods Express* (S. Hughes and R. Lasken, eds.)
© Scion Publishing Limited, 2005

2. METHODS AND APPROACHES

A genome representation method based on adaptor-ligation-mediated PCR has been used for a number of genetic analyses (9–15), including measuring relative gene copy number by Southern blotting or CGH. However, only 12–24% of the human genome can be covered using traditional methods (12). To improve genome coverage, we have used randomly fragmented DNAs as template, rather than the standard enzymatically generated fragments. The theory behind adopting this approach was that whole genomes of both prokaryotic and eukaryotic organisms have in the past been sequenced using subclones prepared by random fragmentation of bacterial artificial chromosome clones or genomic DNA.

This chapter describes a method of WGA, termed adaptor-ligation PCR of randomly sheared genomic DNA (PRSG) (16), that provides a faithful means for amplification of the whole genome. In addition, since this approach does not have a requirement for high-molecular-weight DNA, it is ideally suited for amplification of DNA extracted from methanol-fixed, paraffin-embedded (MFPE) or formalin-fixed, paraffin-embedded (FFPE) material.

3. RECOMMENDED PROTOCOLS

PRSG consists of three stages:

1. A reproducible hydrodynamic shearing of genomic DNA using an automated instrument, HydroShear (see *Protocol 1*).
2. BAL31 nuclease treatment of the sheared DNA (see *Protocol 2*), end filling with T4 DNA polymerase (see *Protocol 3*) and adaptor ligation (see *Protocol 4*).
3. Two steps of PCR with low cycle numbers.

A flow diagram of PRSG is shown in *Fig. 1*. The DNA used in these experiments can be obtained from whole blood, cultured cells, frozen tissue, or laser-capture microdissected tissue.

Protocol 1

Preparation of high-molecular-weight genomic DNA for PRSG

Equipment and Reagents
- HydroShear machine (GeneMachines)
- Phenol
- Chloroform
- Glycogen (Invitrogen)
- 7.5 M Ammonium acetate
- Isopropanol (2-propanol)
- Ethanol
- TE buffer (10 mM Tris-HCl (pH 7.5); 1 mM EDTA)

Method
1. Shear 1 μg of high-molecular-weight genomic DNA using an automated hydrodynamic shearing machine (such as the HydroShear machine), according to the supplier's instructions (17)[a,b].

2. Add an equal volume of phenol to the solution of randomly fragmented DNA and mix for 5 min on a rotating platform.

3. Centrifuge the samples for 10 min at 12 000 g.

4. Transfer the upper aqueous layer to a fresh tube and add an equal volume of chloroform. Mix for 5 min on a rotating platform.

5. Centrifuge for 10 min at 12 000 g.

6. Transfer the upper aqueous layer to a fresh tube and add 1 μl of glycogen (20 μg), 0.5 volumes of 7.5 M ammonium acetate and 2.5 volumes of 100% 2-propanol. Mix well and incubate the solution at room temperature for 20 min.

7. Centrifuge for 10 min at 12 000 g to pellet the DNA.

8. Remove the supernatant and wash the pellet with 1 ml of 70% ethanol. Allow the DNA pellet to air dry.

9. Dissolve the pellet in 10 μl of TE buffer.

Notes
[a]The Hydroshear machine uses a ruby with a 0.05 mm diameter hole to shear the DNA, an approach specific to this piece of equipment. We have not tried any other approaches in our laboratory, but alternative machines or methods that generate DNA fragments within the desired size range (0.5–2 kb) will probably yield comparable results.

[b]When using the Hydroshear machine, the DNA solution (200 μl) was randomly fragmented at appropriate flow rates (speed codes (s.c.) 4 or 5) for 20 iterations.

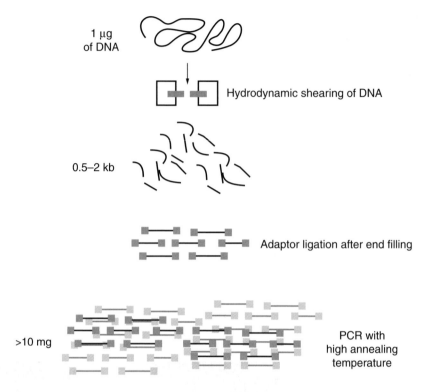

Figure 1. Flow diagram showing the three stages of PRSG.

After treatment with the Hydroshear machine, the sheared DNA fragments (>90%) should fall within a fourfold size distribution ranging from 0.5 to 2 kb in size (see *Fig. 2a*). The DNA is now ready for BAL31 treatment (see *Protocol 2*).

Protocol 2

BAL31 treatment of DNA

Equipment and Reagents
- BAL31 nuclease (1.5 units/µl) and the accompanying BAL31 reaction buffer (Takara Bio Inc.)
- TE buffer (10 mM Tris-HCl (pH 7.5); 1 mM EDTA)

Method
1. Mix 5 µl (approximately 500 ng) of DNA solution from *Protocol 1* with 50 µl of BAL31 reaction buffer and incubate at 70°C for 5 min, followed by 5 min at 30°C.

2. Add 1 µl of BAL31 nuclease and incubate at 30°C for 1 min only.

3. Purify the DNA fragments by phenol extraction followed by precipitation in 2-propanol with 20 µg glycogen, as described in *Protocol 1*, steps 2–8.

4. Dissolve the pellet in 8 µl of TE buffer.

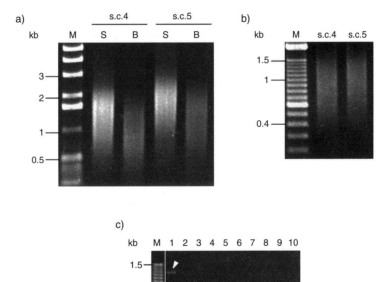

Figure 2. Ethidium bromide-stained gels showing the various stages of PRSG.
(*a*) One microgram of sheared DNA (S) using the HydroShear machine at two different flow rates (speed codes (s.c.) 4 and 5) and the corresponding blunt-ended DNA (B) following BAL31 treatment. (*b*) PRSG amplification products. (*c*) PCR products for several of the exons from the adenomatous polyposis of the colon (*APC*) gene amplified from 50 ng of PRSG DNA. *APC*-specific PCRs were performed using previously reported primers (23, 24). The expected products from the PCR are indicated by arrowheads.

Protocol 3

End filling

Equipment and Reagents
- T4 DNA polymerase (1 unit/μl) and the accompanying 10× T4 polymerase buffer (Takara Bio Inc.)
- TE buffer (10 mM Tris-HCl (pH 7.5); 1 mM EDTA)

Method
1. Add 1 μl of 10× T4 DNA polymerase buffer to the DNA from *Protocol 2* (in 8 μl of TE buffer). Incubate the solution at 70°C for 5 min and then at 30°C for 5 min.

2. Add 1 μl of T4 DNA polymerase and incubate at 37°C for 5 min.

3. Purify the DNA fragments by phenol extraction followed by 2-propanol precipitation (see *Protocol 1*, steps 2–8).

4. Dissolve the pellet in 25 μl of TE.

To amplify the DNA it is necessary to first ligate the *Eco*RI–*Not*I–*Bam*HI adaptors on to each end of the fragmented DNA molecules, as described in *Protocol 4*. The *Eco*RI–*Not*I–*Bam*HI adaptor has the following sequence:

5′-HO-AATTCGGCGGCCGCGGATCC-3′
3′-GCCGCCGGCGCCTAGG-5′

Protocol 4

Ligation of adaptors and PCR amplification

Equipment and Reagents
- *Eco*RI–*Not*I–*Bam*HI adaptor (10 pmol/μl) (Takara Bio Inc.)
- T4 DNA ligase (350 units/μl) and the accompanying 10× T4 DNA ligase buffer (Takara Bio Inc.)
- DNase-free water (autoclaved, deionized)
- 10 mM ATP
- ER-1 PCR primer (5′-GGAATTCGGCGGCCGCGGATCC-3′) (100 pmol/μl)
- 10× PCR buffer (100 mM Tris-HCl (pH 7.5); 500 mM KCl; 30 mM MgCl$_2$)
- 2.5 mM dNTP mix (all four dNTPs)
- ExTaq DNA polymerase (2.5 units/μl) (Takara Bio Inc.)[a]
- TE buffer (10 mM Tris-HCl (pH 7.5); 1 mM EDTA)

Method
1. Combine 1 μl of the DNA solution from *Protocol 3* (20 ng) with 2 μl of 10× T4 DNA ligase reaction buffer, 1 μl of *Eco*RI–*Not*I–*Bam*HI adaptor, 1 μl of 10 mM ATP, 14 μl of DNase-free water and 1 μl of T4 DNA ligase. Incubate at 16°C for 12 h[b].

2. For PCR amplification, combine 1 μl (1 ng) of the adaptor-ligated DNA mixture, 77 μl of DNase-free water, 10 μl of 10× PCR buffer, 1 μl of ER-1 primer, 10 μl of 2.5 mM dNTP mix and 1 μl ExTaq DNA polymerase.

3. Use the following PCR profile for amplification: 15–20 cycles of denaturation at 94°C for 1 min and annealing/extension at 72°C for 3 min, followed by incubation at 72°C for 10 min[c].

4. Aliquot the sample into five separate tubes (20 μl each) and add 58 μl of DNase-free water, 10 μl of 10× PCR reaction buffer, 1 μl of ER-1 primer, 10 μl of 2.5 mM dNTP mix and 1 μl of ExTaq DNA polymerase.

5. PCR amplify each aliquot for an additional five to ten cycles, using the profile described in step 3 of this protocol[d].

6. Purify the DNA fragments by phenol extraction followed by 2-propanol precipitation (see *Protocol 1*, steps 2–8)[e].

7. Dissolve the DNA pellet in 100 μl of TE buffer[f].

Notes

[a]ExTaq DNA polymerase is a hot-start DNA polymerase. Although we have not tried or tested any other hot-start enzymes, we would expect that the results obtained would be similar to those presented here.

[b]If not using commercially obtained adaptors, 'in-house' adaptor molecules can be synthesized as described in *Protocol 5*. When using 'in-house' adaptors, modify step 1 of *Protocol 4* as follows. Combine 1 μl of the DNA solution from *Protocol 3* (20 ng) with 2 μl of 10× T4 DNA ligase buffer, 3 μl of 'in-house' adaptor, 1 μl of 10 mM ATP, 12 μl of DNase-free water and 1 μl of T4 DNA ligase (350 units), and then incubate at 16°C for 12 h. The rest of the procedure is then carried out as described in *Protocol 4*.

[c]Some DNA fragments have a high GC content and may form a stable secondary structure, which often prevents PCR amplification at a standard annealing temperature of 50–60°C. However, by using an adaptor in conjunction with a high annealing temperature (72°C), the amplification bias during PCR due to differences in template sequence composition can be minimized. This in turn allows better genome coverage. We have also developed a modification of the adaptor-ligation-mediated PCR technique for high-fidelity global mRNA amplification for *in vivo* gene expression profiling from as few as 100 microdissected cells (18).

[d]To obtain sufficient yields of DNA, using a high-cycle-number PCR is not recommended as this may introduce sequence bias caused by mis-priming and preferential amplification of shorter fragments. This is why we suggest splitting the first PCR five ways after the initial 15–20 cycles of PCR and then performing an additional five to ten cycles on each aliquot.

[e]We have not used or tested any DNA purification kits, but they may be a suitable substitute for phenol extraction. However, we cannot recommend any particular kit.

[f]From 1 ng of DNA template, PRSG can generate 5–10 μg of amplified product ranging in size from 0.4 to 1.5 kb when the initial HydroShear flow rate is set at s.c. 4, or from 0.5 to 2.0 kb when set at s.c. 5 (see *Fig. 2a*). As an alternative to ExTaq DNA polymerase, we have also tried a high-fidelity enzyme, KOD-Plus DNA polymerase. However, this generated only 2–5 μg of amplified DNA ranging from 0.4 to 1.0 kb at s.c. 4 or from 0.5 to 1.2 kb at s.c. 5 (16).

The adaptor required for *Protocol 4* may be synthesized in the laboratory rather than obtained commercially; a suitable method is given in *Protocol 5*.

Protocol 5

Construction of adaptor molecule

Equipment and Reagents
- Oligonucleotide 1 (5′-AATTCGGCGGCCGCGGATCC-3′) (100 pmol/µl)
- Oligonucleotide 2 (3′-GCCGCCGGCGCCTAGG-5′) (100 pmol/µl)
- 10× T4 DNA ligase buffer (Takara Bio Inc.)
- DNase-free water (autoclaved, deionized)

Method
1. Combine 0.3 µl of 10× T4 DNA ligase buffer, 0.5 µl of oligonucleotide 1, 0.5 µl of oligonucleotide 2, and 1.7 µl DNase-free water.

2. Incubate the sample using a programmed temperature gradient in a thermocycler from 65 to 15°C ramping at –1°C/min to form the adaptor complexes.

Protocol 6

PRSG for laser-capture microdissected samples[a]

Equipment and Reagents
- Lysis buffer (10 mM Tris-HCl (pH 7.5); 1 mM EDTA; 0.5% SDS)
- Proteinase K (10 mg/ml)
- TE buffer (10 mM Tris-HCl (pH 7.5); 1 mM EDTA)

Method
1. Microdissect the regions of interest following the guidelines of the relevant equipment manufacturers[b].

2. To extract DNA, mix the dissected cells (at least 1000–2000 cells) with 200 µl of lysis buffer.

3. Add 2 µl of proteinase K and incubate the reaction for 15 h at 42°C.

4. Purify the DNA by phenol extraction followed by 2-propanol precipitation (see *Protocol 1*, steps 2–8). It is unlikely that a DNA pellet will be visible, so it is important to decant the supernatant carefully so as not to accidentally discard the DNA.

5. Dissolve the DNA pellet in 10 µl of TE buffer.

6. Amplify by adaptor-ligation-mediated PCR after end filling as described in *Protocols 2–4*.

Notes
[a]We used MFPE carcinoma tissue sections of 8 µm for laser microdissection.
[b]We have used the PixCell II LCM system (Arcturus Engineering), PALM MicroLaser Systems (PALM Microlaser Technologies AG) and Laser Microdissection System (Molecular Machines & Industries).

We have also obtained DNA for PRSG from MFPE esophageal carcinoma tissues (19). This requires only slight modification to the above protocols and is described in *Protocol 6*. From 100–1000 captured cancer cells, we typically obtain 1–10 ng of DNA, from which PRSG can generate hundreds of micrograms of DNA. It is not necessary to perform hydrodynamic shearing as the DNA is already fragmented as a result of the fixation process.

3.1. Downstream applications

3.1.1. Analysis of unique genomic DNA sequences

An assessment of the genome representation provided by PRSG was carried out by performing PCRs on more than 2000 exons randomly distributed throughout the genome. These PCRs were performed on PRSG products generated from high-quality DNA from 400 individuals. A failure rate of less than 1% was seen for PCRs using PRSG-amplified DNA as template. As an example, the PCR results for ten exons of the *APC* gene are shown in *Fig. 2c*. However, the PCR failure rate using PRSG-amplified DNA from FFPE tissues was sample dependent and ranged from 30 to 50%, suggesting that PRSG may not be the best approach for generating such DNA templates for PCR. Nevertheless, PRSG products from FFPE tissues have been successfully used for array CGH analysis (16).

3.1.2. Single nucleotide polymorphism (SNP) analysis

We performed direct sequence analysis of the PCR products to examine whether the PRSG products were useful for SNP detection. In order to verify the conservation of 12 SNPs within the PRSG products, parallel genotyping was performed with both the PRSG products and the corresponding original genomic DNA of 48 individuals. Using pyrosequencing (20), identical results were obtained for the PRSG products and the corresponding original genomic DNAs (16). These results suggest that PRSG-amplified DNA may be of use for large-scale SNP scoring studies.

3.1.3. Microsatellite analysis

Microsatellite analysis was performed on PRSG-amplified DNA and matched non-amplified DNA. Of 307 microsatellites distributed across the genome, 258 (84%) were reproducibly amplified from both the PRSG product and the original DNA, whilst 49 (16%) of the microsatellites were not present in the PRSG-amplified DNA. This finding was observed in repeated experiments using PRSG-amplified DNA. Of the 258 microsatellites, 256 (99%), including 120/122 heterozygotes and 136/136 homozygotes, showed a consistent pattern between the PRSG product and the original DNA. These results indicate that the majority of microsatellites are retained in PRSG-amplified DNA (16).

a)   Southern blot analysis

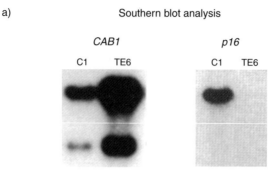

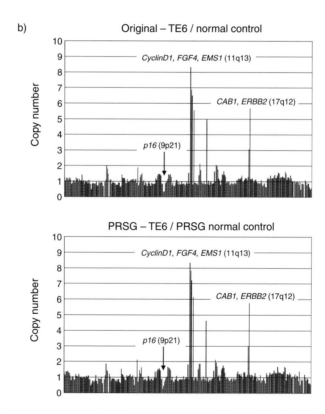

Figure 3. Relative gene copy number as detected by Southern blot analysis and array CGH using PRSG-amplified and the corresponding non-amplified genomic DNA.

(*a*) Southern blot data showing proliferation differentiation and control protein 1 (*CAB1*) amplification and cyclin-dependent kinase inhibitor 2A (*p16*) homozygous deletion in the TE6 esophageal cancer cell line compared with normal esophageal control DNA (C1). (*b*) Representative results of array CGH on PRSG products (lower panel) compared with the original non-amplified genomic DNA (upper panel). In each case, the data indicate the copy number of specific genes when comparing DNA from TE6 cells and normal esophageal control DNA. Genes that show an alteration in copy number are indicated.

3.1.4. Southern blotting and array CGH analysis

In the TE6 esophageal cancer cell line, amplification of the proliferation differentiation and control protein 1 (*NPDC1/CAB1*), avian erythroblastic leukemia viral oncogene homolog 2 (*ERBB-2*), CyclinD1/CCND1 and fibroblast growth factor 4 (*FGF4*) genes and the oncogene *EMS1* have been reported, as well as deletion of the cyclin-dependent kinase inhibitor 2A (*CDKN2A/p16*) gene (21, 22). To evaluate the locus representation of PRSG-amplified DNA, we conducted Southern blotting and array CGH analysis (287 chromosomal loci). Southern blot analysis of PRSG-amplified normal esophageal control DNA (C1) and TE6 DNA demonstrated the expected amplification of *CAB1* and loss of *p16* (see *Fig. 3a*). CGH experiments performed using TE6 DNA vs. normal esophageal control DNA showed more than 90% concordance of the fluorescence ratios between the PRSG-amplified and matched non-amplified DNA, demonstrating that the locus representation in the PRSG product is comparable to that in the original genome DNA (*Fig. 3b*).

4. TROUBLESHOOTING

- In the automated hydrodynamic shearing (*Protocol 1*), it is necessary to avoid using samples containing more than 5 µg of DNA, insoluble DNA, or other contaminants since the small hole in the ruby of the HydroShear machine is easily blocked.
- When using an alternative approach for shearing DNA other than the HydroShear machine, we suggest that the protocol first be performed using a DNA sample for which results are known.
- To amplify fragmented DNA effectively, the quantity of template DNA and the $MgCl_2$ concentration are critical; 0.5–2 ng of DNA (or 1000–2000 captured cells) should be used as template in a PCR tube containing 2–3 mM $MgCl_2$. In our experience, higher amounts of DNA cause suboptimal PCR amplification, the reason for which is unknown.
- Of all the processes involved in PRSG, only DNA shearing is relatively time consuming, and hence the processing of multiple samples is best carried out using an automated system.

5. REFERENCES

★ 1.   Telenius H, Pelmear AH, Tunnacliffe A, *et al.* (1992) *Genes Chromosomes Cancer,* 4, 257–263. – *First description of DOP-PCR.*
★ 2.   Zhang L, Cui X, Schmitt K, Hubert R, Navidi W & Arnheim N (1992) *Proc. Natl. Acad. Sci. U. S. A.* 89, 5847–5851. – *First description of PEP.*
   3.   Snabes MC, Chong SS, Subramanian SB, Kristjansson K, Disepio D & Hughes MR (1994) *Proc. Natl. Acad. Sci. U. S. A.* 91, 6181–6185.
   4.   Cheung VG & Nelson SF (1996) *Proc. Natl. Acad. Sci. U. S. A.* 93, 14676–14679.
★ 5.   Dean FB, Hosono S, Fang L, *et al.* (2002) *Proc. Natl. Acad. Sci. U. S. A.* 99, 5261–5266. – *First description of strand-displacement amplification.*

6.   Kuukasjarvi T, Tanner M, Pennanen S, Karhu R, Visakorpi T & Isora J (1997) *Genes Chromosomes Cancer*, **18**, 94–101.
7.   Dietmaier W, Hartmann A, Wallinger S, *et al.* (1999) *Am. J. Pathol.* **154**, 83–95.
8.   Huang Q, Schantz SP, Rao PH, Mo J, McCormick SA & Chaganti RSK (2000) *Genes Chromosomes Cancer*, **28**, 395–403.
9.   Ko MSH, Ko SBH, Takahashi N, Nishiguchi K & Abe K (1990) *Nucleic Acids Res.* **18**, 4293–4294.
10. Sasaki H, Nomura S, Akiyama N, *et al.* (1994) *Cancer Res.* **54**, 5821–5823.
11. Inoue S, Kiyama R & Oishi M (1996) *Genomics*, **31**, 271–276.
★ 12. Lucito R, Nakimura M, West JA, *et al.* (1998) *Proc. Natl. Acad. Sci. U. S. A.* **95**, 4487–4492. *– Description of the use of 'genomic representations' for genetic analysis.*
13. Ohki R, Oishi M & Kiyama R (1998) *Mol. Carcinog.* **22**, 158–166.
★ 14. Klein CA, Schmidt-Kitteler O, Schardt JA, Pantel K, Speicher MR & Riethmuller G (1999) *Proc. Natl. Acad. Sci. U. S. A.* **96**, 4494–4499. *– A report on the use of ligation-mediated PCR for amplification of DNA from a single cell.*
15. Makrigiorgos GM, Chakrabarti S, Zhang Y, Kaur M & Price BD (2002) *Nat. Biotechnol.* **20**, 936–939.
★★★ 16. Tanabe C, Aoyagi K, Sakiyama T, *et al.* (2003) *Genes Chromosomes Cancer*, **38**, 168–176. *– First description of the PRSG method.*
17. Thorstenson YR, Hunicke-Smith SP, Oefner PJ & Davis RW (1998) *Genome Res.* **8**, 848–855.
18. Aoyagi K, Tatsuta T, Nishigaki M, *et al.* (2003) *Biochem. Biophys. Res. Commun.* **300**, 915–920.
19. Emmert-Buck MR, Bonner RF, Smith PD, *et al.* (1996) *Science*, **274**, 998–1001.
20. Alderborn A, Kristofferson A & Hammerling U (2000) *Genome Res.* **10**, 1249–1258.
21. Igaki H, Sasaki H, Tachimori Y, *et al.* (1995) *Cancer Res.* **55**, 3421–3423.
22. Ishizuka T, Tanabe C, Sakamoto H, *et al.* (2002) *Biochem. Biophys. Res. Commun.* **296**, 152–155.
23. Groden J, Thliveris A, Samowitz W, *et al.* (1991) *Cell*, **66**, 589–600.
24. Yamakita N, Murai T, Ito Y, *et al.* (1997) *Intern. Medicine*, **36**, 536– 542.

CHAPTER 6

GenomePlex whole genome amplification

Simon Hughes[1], Gabrielle Sellick[1], Richard Coleman[1] and John Langmore[2]

[1]Institute of Cancer Research, Section of Cancer Genetics, 15 Cotswold Road, Surrey, UK; [2]Rubicon Genomics, 4370 Varsity Drive, Suite G, Ann Arbor, Michigan, USA

1. INTRODUCTION

The availability of sufficient quantities of genomic DNA is crucial for numerous assays used in the study of human and animal disease. However, acquisition of this most basic of requirements is often the stumbling block for most researchers. In many cases, patient material (histological sections, tissue samples, buccal cells, or peripheral blood) is limited and when studying tissue sections, particular regions of interest may only consist of a few hundred cells. As a result the yield of DNA is often insufficient for extensive analysis and, once exhausted, is impossible to replace. As a consequence, the total number of experiments that can be performed is often restricted to region-specific approaches, in contrast to more informative whole genome screens. Although the former approach may be successful (1, 2), it is possible that important genetic information essential for making prognostic and/or therapeutic decisions may go unnoticed (3). To address these problems, several WGA techniques have been developed (4–15) that are capable of amplifying genomic DNA. A number of these methods are described elsewhere in this book. Depending on the approach used, several micrograms of DNA can be generated from as little as 1 ng of starting material. The advantages of WGA are that it provides sufficient DNA for numerous assays to be performed immediately, as well as providing DNA that can be archived for future investigation. Progress in the field of WGA, which started in the early 1990s, has greatly eased the constraints imposed by limited genomic DNA. As a result, the amount of data that can be derived from limited clinical samples has significantly increased, as has the prospect of expanding our knowledge of human diseases, where the quantity of DNA was initially a limiting factor.

Whole Genome Amplification: *Methods Express* (S. Hughes and R. Lasken, eds.)
© Scion Publishing Limited, 2005

2. METHODS AND APPROACHES

The choice of WGA methodology is often dependent on the source of tissue from which DNA will be extracted. For instance, strand-displacement amplification (4, 16–22) has a requirement for high-quality, high-molecular-weight DNA, usually in excess of 2 kb (see Chapter 8 by Lasken), which can often only be obtained from fresh tissue or blood and not from fixed tissue. In contrast, PCR-based methods (6, 8–10) are generally less affected by DNA quality and are more applicable to DNA extracted from various sources (fixed and fresh tissue). When implementing WGA in the laboratory, it is important to assess the entire experimental process closely, including sample collection, fixation, storage, and initial DNA extraction procedures, as all of these factors can affect DNA quality and thus have some bearing on the selection of WGA technique. When using WGA, it is important first to validate the method selected and to become proficient in the technique before applying it to clinical samples. Irrespective of the method of choice, it is essential that the results generated from the amplified DNA are indistinguishable from the results obtained from the original genomic DNA.

This chapter will provide an overview of the GenomePlex WGA technology (a PCR-based WGA method) as used in our laboratory. The results presented here demonstrate its use for generating large quantities of DNA. As an example of its utility, typical results obtained for PCR, sequencing, microsatellite analysis, and comparative genomic hybridization (CGH) will be presented.

2.1. GenomePlex WGA

The GenomePlex WGA technology (see *Fig. 1*) was developed as a universal tool to produce sufficient quantities of DNA for high-throughput analysis and for the generation of banked DNA for future genetic tests. This method utilizes a proprietary amplification technology that is based upon random fragmentation of genomic DNA. The process begins by converting the complete genome into an *in vitro* molecular library of DNA fragments of controlled length by random, nonenzymatic fragmentation. This is followed by an hour-long incubation at varying temperatures, which facilitates the addition of adaptor sequences, containing PCR priming sites, to both ends of every fragment (7, 23). This fragment library can then be amplified several 1000-fold by PCR, potentially generating milligram amounts of DNA from starting concentrations of 10–100 ng. Furthermore, aliquots of the amplified library can be re-amplified to achieve in excess of a 100 000-fold amplification. Library preparation and amplification generally take less than 4 h and, due to the limited number of experimental steps, lend themselves to automation.

2.1.1. DNA source

The source of DNA that can be accurately and robustly amplified using the GenomePlex protocol is almost unlimited and includes DNA extracted from fixed, frozen, or archival tissue, whole blood, buccal swabs, single cells, sorted

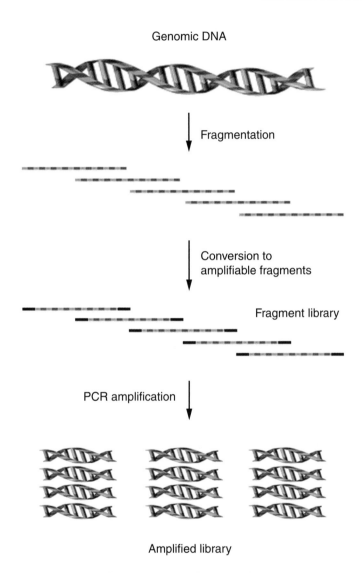

Figure 1. Graphical representation of GenomePlex WGA.
DsDNA is first denatured and fragmented by heating at 95°C for 4 min in the presence of Fragmentation Buffer. Library Preparation Buffer and Library Stabilization Solution are then mixed with the DNA fragments and heated at 95°C for 2 min. The DNA fragments are converted to an amplifiable library by the ligation of adaptors to the fragment ends. This is facilitated by the addition of Library Preparation Enzyme and a step-wise thermal cycling program of 16°C for 20 min, 24°C for 20 min, 37°C for 20 min and 75°C for 5 min. The fragment library can then be amplified by PCR using the Amplification Master Mix and either JumpStart or BD TITANIUM *Taq* DNA polymerase.

chromosomes, and laser-capture microdissected cells. The amplified DNA produced is suitable for a range of downstream genetic assays (24–26) and thus has the potential for use not only in academic research, but also in commercial, forensic, and diagnostic laboratories.

3. RECOMMENDED PROTOCOLS

A flow diagram of the methods involved is shown in *Fig. 2.*

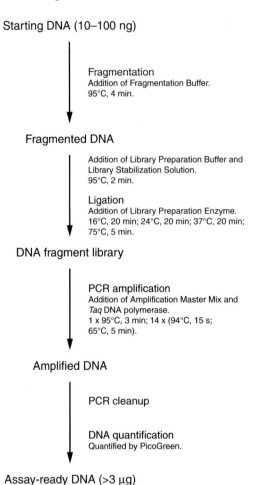

Starting DNA (10–100 ng)

Fragmentation
Addition of Fragmentation Buffer.
95°C, 4 min.

Fragmented DNA

Addition of Library Preparation Buffer and
Library Stabilization Solution.
95°C, 2 min.

Ligation
Addition of Library Preparation Enzyme.
16°C, 20 min; 24°C, 20 min; 37°C, 20 min;
75°C, 5 min.

DNA fragment library

PCR amplification
Addition of Amplification Master Mix and
Taq DNA polymerase.
1 x 95°C, 3 min; 14 x (94°C, 15 s;
65°C, 5 min).

Amplified DNA

PCR cleanup

DNA quantification
Quantified by PicoGreen.

Assay-ready DNA (>3 μg)

Figure 2. Step-by-step quick reference guide to GenomePlex WGA.

Protocol 1

DNA extraction and quantification

Equipment and Reagents

- QIAamp DNA Blood Midi Kit (Qiagen)
- QIAamp DNA Blood Mini Kit (Qiagen)
- 1% Agarose gel containing 10 ng/ml ethidium bromide
- 6× Orange loading dye solution (Fermentas)
- Equipment and reagents for agarose gel electrophoresis including 1× TBE agarose gel running buffer (10.8 g/l Tris base; 5.5 g/l boric acid; 4 ml/l 0.5 M EDTA, pH 8.0; diluted from a 10× stock) (Sigma)
- UV light source
- RediPlate 96 PicoGreen dsDNA Quantitation Kit (Invitrogen)
- Fluorescence-based microplate readers or fluorometer

Method

1. Extract DNA from up to 2 ml of whole blood using the QIAamp DNA Blood Midi Kit, following the manufacturer's instructions[a,b].

2. Extract DNA from tissue sections (four sections of 5 μm thickness) using the QIAamp DNA Blood Mini Kit, following the manufacturer's instructions.

3. Determine the size of the DNA by analyzing a 5 μl aliquot by agarose gel electrophoresis (1% agarose gel containing 10 ng/ml of ethidium bromide). This will also assess the degree of DNA degradation. Detect the DNA under UV light.

4. Dilute the DNA extracted from blood 1:4000 in sterile water and from tissue sections 1:1000 in sterile water for quantification.

5. Quantify the DNA concentration using the RediPlate 96 PicoGreen dsDNA Quantitation Kit (or similar kit) in conjunction with a fluorescence-based microplate reader[c], following the manufacturer's instructions.

6. Prepare DNA solutions at a concentration of 1 ng/μl for DNA from fresh tissue or cells. When using DNA extracted from fixed tissue, prepare the DNA solution at a concentration of 10 ng/μl.

Notes

[a]GenomePlex will amplify DNA from a variety of sources including buffy coats, buccal swabs, cultured cells, blood spots, and mouthwashes.

[b]Fixation of tissues can introduce sequence variations and reduce overall DNA quality. When studying such tissue, prior examination of the DNA by agarose gel electrophoresis will help determine DNA quality.

[c]DNA concentration (μg/μl) can also be quantified using a standard spectrophotometer by taking the absorbance reading at 260 nm and multiplying it by 50 and then by the dilution factor.

Protocol 2

Fragmentation

Equipment and Reagents
- 0.2 ml Flat-cap tubes (ABgene)
- Strips of eight Thermo-Tubes and flat-cap strips (ABgene)
- Thermo-Fast 96 Semi-Skirted PCR Plate (ABgene)
- GenomePlex Whole Genome Amplification Kit (Sigma)[a]
- Thermal cycler

Method
1. Combine 10 µl of DNA sample (final concentration 10–100 ng) with 1 µl of 10× Fragmentation Buffer (blue-capped tube)[b] from the GenomePlex Whole Genome Amplification Kit in a 0.2 ml flat-cap tube or a multi-well strip/96-well PCR plate[c,d].

2. Mix the sample by either pipetting or brief vortexing.

3. Consolidate the sample by centrifugation (5–10 s).

4. Incubate at 95°C for 4 min in a thermal cycler[e].

5. Following incubation, cool the sample on ice for 5 min.

Notes

[a]The GenomePlex Kit was originally developed by Rubicon Genomics, but is now available from Sigma. The only difference is the suggested *Taq* DNA polymerase used for the amplification step.

[b]The constituents of the GenomePlex Kit buffers are proprietary and therefore unknown.

[c]When handling many samples (>20), we suggest using multi-well strips or 96-well PCR plates, as this will help decrease the set-up time. However, caution must be taken when removing the strip caps to avoid cross contamination of tube contents. For 96-well plates, we suggest the use of adhesive metal or plastic films. The advantage of these over strip caps is that they do not need to be removed and can be pierced using a pipette tip to allow the addition of the library preparation solutions. The plate can then be resealed by placing a second film over the top of the first film.

[d]We would suggest users prepare at least two GenomePlex reactions for each sample as this will provide a greater yield of DNA. In addition, we have also obtained better results in downstream applications when the products from at least two reactions are combined (see section 4 on Troubleshooting).

[e]As stated by Sigma in the manual for the GenomePlex Kit, adhering to this incubation time is essential as longer or shorter times can affect results. Although we have not tried to alter this step, when using degraded DNA decreasing the fragmentation time may improve results.

Protocol 3

Library preparation

Equipment and Reagents
■ GenomePlex Whole Genome Amplification Kit (Sigma)
■ Thermal cycler

Method

1. Add 2 µl of 1× Library Preparation Buffer[a] (green-capped tube) and 1 µl of Library Stabilization Solution[a] (yellow-capped tube) to each sample.

2. Mix the sample by either pipetting or brief vortexing.

3. Consolidate the sample by centrifugation (5–10 s).

4. Incubate at 95°C for 2 min in a thermal cycler.

5. Following incubation, cool the sample on ice for 5 min and consolidate by centrifugation (5–10 s).

6. Add 1 µl of Library Preparation Enzyme[a] (orange-capped tube), mix by pipetting or vortexing, and centrifuge briefly.

7. Incubate the samples in a thermal cycler using the following conditions: 16°C for 20 min, 24°C for 20 min, 37°C for 20 min, and 75°C for 5 min[b].

8. Store the reaction mixtures at –20°C for up to 3 days or continue with PCR amplification (see *Protocol 4*).

Notes

[a]From the GenomePlex Whole Genome Amplification Kit.
[b]Following library preparation, samples can be stored at –20°C for up to 3 days. We have not stored them for longer than this, so cannot comment on the effect of long-term storage on WGA DNA and subsequent downstream applications.

Protocol 4

PCR amplification protocol

Equipment and Reagents
- JumpStart *Taq* DNA Polymerase (Sigma) or BD TITANIUM *Taq* DNA Polymerase (BD Clontech).
- 100 bp DNA ladder (Invitrogen)
- 1% Agarose gel containing 10 ng/ml ethidium bromide
- 6× Orange loading dye solution (Fermentas)
- Equipment and reagents for agarose gel electrophoresis including 1× TBE agarose gel running buffer (Sigma)

Method

1. The volume of reaction constituents for this stage is dependent on the source of *Taq* DNA polymerase being used: JumpStart or BD TITANIUM. Both enzymes give comparable DNA yields[a].

   Per reaction, combine 7.5 μl of 10× Amplification Master Mix[b] (red-capped tube), 47.5 μl of sterile water, 5 μl of JumpStart *Taq* DNA polymerase and the 15 μl from the library preparation step (see *Protocol 3*, step 8).

   or

   Per reaction, combine 7.5 μl of 10× Amplification Master Mix[b] (red-capped tube), 51.75 μl of sterile water, 0.75 μl of BD TITANIUM *Taq* DNA polymerase and the 15 μl from the library preparation step (see *Protocol 3*, step 8).

2. Mix the reaction constituents thoroughly by pipetting or vortexing, and centrifuge briefly.

3. Use the following PCR profile for amplification: initial denaturation at 95°C for 3 min, followed by 14 cycles of denaturation at 94°C for 15 s and annealing/extension at 65°C for 5 min[c].

4. Determine the size of the product by mixing 5 μl of the reaction mix with 1 μl of 6× orange loading dye solution and resolving the aliquot by agarose gel electrophoresis (1% agarose gel containing 10 ng/ml of ethidium bromide) alongside 4 μl of a molecular weight marker (100 bp DNA ladder)[d] (see *Fig. 3*).

5. Detect the DNA smears under UV light.

6. Store the reaction mixtures at −20°C prior to purification.

Notes

[a]We strongly recommend using either the JumpStart or BD TITANIUM *Taq* DNA polymerase, as both have been optimized, by either Sigma (JumpStart *Taq* DNA polymerase) or Rubicon Genomics (BD TITANIUM *Taq* DNA polymerase), for use with the GenomePlex Kit. We cannot guarantee satisfactory results with other sources of *Taq* DNA polymerase.

[b]From the GenomePlex Whole Genome Amplification Kit.

[c]If more than one reaction has been set up for each sample (which we recommend), combine reactions at this point.

[d]The size of the amplification product is dependent on the quality of the starting DNA. If this was high-molecular-weight DNA (extracted from fresh tissue or cells), the amplification product smear will range from 50 to 2000 bp. However, if the DNA was of low molecular weight (from fixed tissue), then the size of the amplification product smear will generally be below 500 bp.

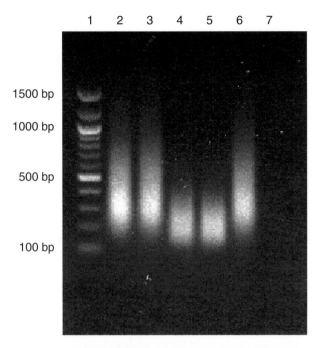

Figure 3. Example of the DNA smears produced from fresh tissue/blood (50–2000 bp) or fixed tissue (<500 bp).
(*Lane 1*) 100 bp DNA size ladder; (*lanes 2 and 3*) amplification products obtained using DNA extracted from fresh tissue; (*lanes 4 and 5*) amplification products obtained using DNA extracted from fixed tissue; (*lane 6*) positive control (included with kit); (*lane 7*) negative control (no DNA).

An example of the PCR products generated using *Protocol 4* is shown in *Fig. 3*. It is important to clean up GenomePlex WGA products in order to remove unincorporated primers and other reaction constituents that may interfere with downstream applications. We have used the MinElute 96 UF PCR Purification Kit with a vacuum manifold for PCR cleanup, as this allows a high sample throughput. We have also used phenol-chloroform extraction to clean the WGA products, in addition to the QIAquick PCR Purification Kit (Qiagen), DNA Clean & Concentrator 5 (Genetix) and the Microcon YM-30 Centrifugal Filter Unit (Millipore).

Protocol 5

PCR cleanup and quantification

Equipment and Reagents
- MinElute 96 UF PCR Purification Kit (Qiagen)
- Vacuum pump
- Sterile water (Sigma)
- EB buffer (10 mM Tris-HCl (pH 8.5))
- Microplate shaker
- RediPlate 96 PicoGreen dsDNA Quantitation Kit (Invitrogen)
- Fluorescence-based microplate reader or fluorometer

Method

1. Add the amplified product from *Protocol 4* to the top of the membrane from the MinElute Kit and apply the vacuum (~800 millibar) until all wells are completely dry (15–20 min).

2. Add two separate washes[a] of 50 μl of sterile water to each well and apply the vacuum (~800 millibar) until all wells are completely dry (15–20 min).

3. Add 100 μl of EB buffer and shake on a microplate shaker for 5–10 min to dissolve the DNA[b].

4. Transfer the solution to a 1.5 ml tube and store at −20°C until required.

5. Quantify the DNA concentration using the RediPlate 96 PicoGreen dsDNA Quantitation Kit (or similar kit) in conjunction with a fluorescence-based microplate reader, following the manufacturer's instructions[c,d].

Notes

[a]We followed the protocol provided by the manufacturer (Qiagen); however, we recommend performing at least two washes in order to guarantee removal of salts, incorporated nucleotides, primers, etc.

[b]The DNA can also be dissolved by pipetting samples up and down.

[c]PicoGreen quantification can underestimate the amount of DNA generated by WGA due to the generation of ssDNA. As an alternative, the amplified DNA can be quantified by A_{260} using a spectrophotometer. On some occasions, the DNA yield determined by either fluorescence- or spectrophotometer-based quantitation is not in agreement with that observed by resolving the amplification products on a gel (see section 4 on Troubleshooting). Although this has not affected any of our downstream applications, if this is of concern, one possible option is to use real-time PCR to quantify the DNA concentration (27).

[d]We consistently get yields of 2–3 μg per reaction.

3.1. Downstream applications

3.1.1. Region-specific PCR

PCRs are performed following the standard protocol provided with the *Taq* DNA polymerase (ABgene). When using WGA on DNA extracted from blood, we have consistently been able to amplify PCR products ranging from 100 to 600 bp (see *Fig. 4*). We have not used amplified DNA extracted from fixed tissue for PCRs, so cannot comment on how well it will work in amplifying different-sized products. However, we would expect that it would work best when the target PCR amplicons are small (100–300 bp).

3.1.2. Sequence analysis

PCR products generated from GenomePlex-amplified DNA and matched non-amplified DNA were bidirectionally sequenced using BigDye Terminator chemistry (version 3.1) implemented on an ABI 3100 sequencer (Applied Biosystems). When comparing the resultant traces using the SEQUENCHER version 3.1.1 software package (Gene Codes), we have not observed any sequence variations for those amplicons studied (see *Fig. 5*). Furthermore, detailed sequence examination has failed to reveal significant differences in background, peak height, or readable trace length between amplified and non-amplified material.

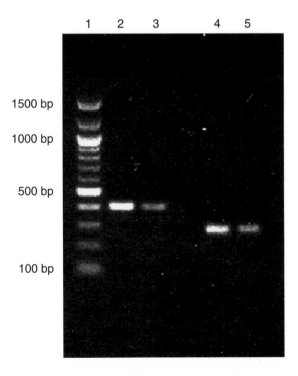

Figure 4. Comparison of the products obtained using non-amplified DNA and GenomePlex-amplified DNA as template for PCR.
(*Lane 1*) 100 bp DNA size ladder; (*lanes 2 and 3*) primer set 1 for non-amplified (*lane 2*) and matched GenomePlex-amplified (*lane 3*) DNA; (*lanes 4 and 5*) primer set 2 for non-amplified (*lane 4*) and matched GenomePlex-amplified (*lane 5*) DNA.

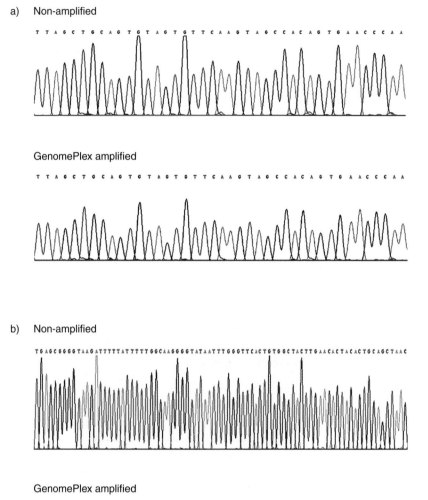

Figure 5. Comparison of the sequence data obtained from non-amplified DNA and matched GenomePlex-amplified DNA for two separate genes.
(*a*) Gene 1; (*b*) gene 2. PCR products were bidirectionally sequenced using BigDye Terminator chemistry (version 3.1) implemented on an ABI 3100 sequencer. Sequence data were analyzed using the SEQUENCHER version 3.1.1 software package.

3.1.3. Microsatellite analysis

Fluorescent microsatellite markers (100–350 bp) have been amplified by PCR from GenomePlex-amplified DNA and matched non-amplified DNA, using standard procedures. Analysis was undertaken on an ABI 3100 genetic analyzer with allele sizes determined by use of the ABI PRISM Genotyper software package (Applied Biosystems). Three types of microsatellite repeats (dinucleotide, trinucleotide, and tetranucleotide) were used for testing. We consistently obtained high-quality results with tri- and tetranucleotide repeats, with GenomePlex-amplified DNA and matched non-amplified DNA giving indistinguishable results. However, the amplification of dinucleotide repeats did not demonstrate concordance between the two DNA sources, so we would suggest that analysis of dinucleotide repeats be avoided when using GenomePlex-amplified DNA (see *Fig. 6*).

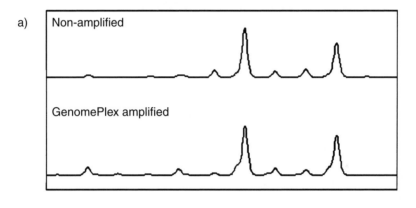

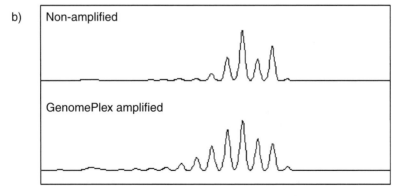

Figure 6. Typical results attained for microsatellite analysis of GenomePlex-amplified and matched non-amplified DNA using primers to amplify (*a*) tetranucleotide repeats and (*b*) dinucleotide repeats.
Fluorescent microsatellite markers (100–350 bp) were amplified by PCR, using standard procedures. Analysis was undertaken on an ABI 3100 genetic analyzer with allele sizes determined by use of the ABI PRISM Genotyper software package.

3.1.4. Array CGH

When using CGH to study copy number imbalances in tumors it is essential to amplify both test and control DNA samples such that any bias introduced by the amplification procedure is equal in both DNAs to be used in the experiment. Array CGH has been performed using a modification of the standard protocol (28). Test (normal male DNA) and control (normal female DNA) DNA was amplified from both commercial sources (Promega, male or female DNA) and from DNA extracted from fixed tissue. Two to three micrograms of amplified DNA was labeled with either Cy5 or Cy3 using the Bioprime Labeling Kit (Invitrogen); when using non-amplified DNA only 0.5–1.5 µg of DNA is generally used. Microarray slides were processed according to standard protocols and analysed using the NORMALISE Suite 2 set of programs (29). Comparisons of the CGH results obtained from non-amplified commercial DNA and GenomePlex-amplified commercial DNA showed no differences, i.e. loss of X and gain of Y in both (see *Fig. 7a* and *b*). This similarity of results was also observed for a number of cell lines (see *Fig. 8*). Analysis of the amplified male DNA versus amplified female DNA (both DNAs extracted from fixed tissue) also showed loss of X and gain of Y (see *Fig. 7c*). The genomic profile in *Fig. 7(c)* shows increased 'noise' represented by a greater deviation from the central line (ratio of 1:1) for the 22 autosomes. This is to be expected, due to the lower quality of the starting DNA, which can cause unequal amplification of genomic regions in the two samples. Although it is unfair to compare directly the results obtained from commercial DNA and from DNA extracted from fixed tissue, this comparison does demonstrate the generally poorer results obtained for amplified DNA from fixed tissue. Despite this weakness, it is possible to detect copy number alterations in non-amplified DNA from fixed tissue (27, 30, 31). Although, this has not yet been reported using GenomePlex WGA DNA, it is likely that satisfactory results will be obtained.

3.1.5. Single nucleotide polymorphism (SNP) genotyping

SNP genotyping has been performed on GenomePlex-amplified DNA obtained from blood and matched non-amplified DNAs using the GeneChip Mapping 10K *Xba* Array (Affymetrix) (32). The signal detection and SNP call rates for the two non-amplified samples were 99.84 and 98.87%, respectively, for sample 1, and 99.78 and 98.55%, respectively, for sample 2. However, the signal detection and SNP call rates for the two GenomePlex-amplified DNAs were much lower: 33.56 and 24.18%, respectively, for sample 1, and 47.32 and 36.70%, respectively, for sample 2. These results indicate that GenomePlex-amplified DNA is not suitable for SNP genotyping when coupled with the Affymetrix experimental protocol. Possibly the reason for this failure is that the random fragmentation used in the GenomePlex WGA does not conserve the *Xba*I restriction sites required for the Affymetrix protocol. However, this remains to be tested.

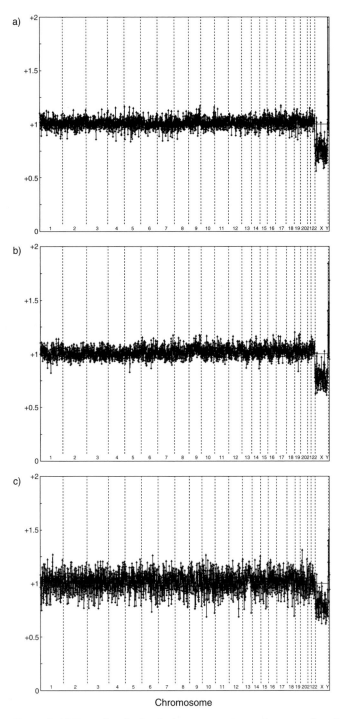

Figure 7. CGH results obtained when comparing male versus female DNA using (*a*) non-amplified commercial (Promega) DNA, (*b*) GenomePlex-amplified commercial (Promega) DNA, and (*c*) GenomePlex-amplified DNA extracted from fixed tissue.
Array CGH was performed using standard protocols for non-amplified DNA. However, minor modifications were required when using GenomePlex-amplified DNA, whereby 2–3 µg of amplified DNA was used rather than the standard 0.5–1 µg of non-amplified DNA.

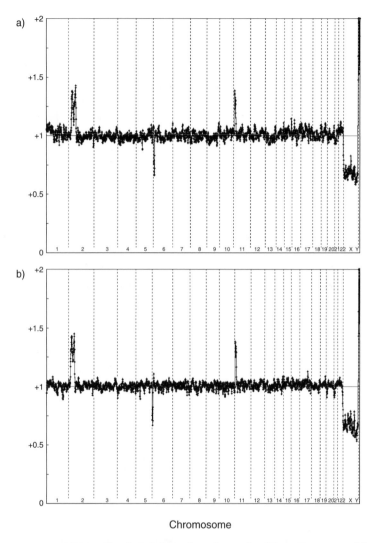

Figure 8. CGH results obtained for the colorectal cell line DLD1 using (*a*) non-amplified DNA and (*b*) amplified DNA.
Gains on chromosome 2 and 11, in addition to loss on chromosome 6, are present in both non-amplified and amplified DNA.

4. TROUBLESHOOTING

- It is important always to include a positive and negative control. The former ensures that the reactions are working optimally and the latter determines whether any of the reaction constituents are contaminated. For the negative control, carry out the procedure as described above, but omit the genomic

DNA. No amplification should be observed in the negative control if the protocol is correctly adhered to.

- The starting concentration of DNA is crucial. Best results are obtained when starting with 10 ng of DNA extracted from fresh tissue/blood or 100 ng of DNA extracted from fixed tissue. Lower amounts of DNA (<10 ng from fresh tissue or <100 ng from fixed tissue) will generate amplification products, but results obtained in downstream applications may not be faithful to the result that would have been obtained from nonamplified DNA.

- Tissue fixation causes degradation of DNA within the sample. Although GenomePlex will amplify DNA larger than 200 bp, it is essential to use increased quantities of starting DNA (100 ng) to guarantee a satisfactory yield of final product.

- The efficiency of amplification is dependent on the quality of the starting DNA. Although GenomePlex is tolerant to mild or moderate DNA degradation, moderate to severe degradation will reduce WGA efficiency. This will result in a decrease in the quality of results obtained in downstream applications.

- We recommend combining at least two individual WGA amplifications for each DNA to be studied, as this has produced better results in downstream experiments. In our CGH studies, when comparing male DNA versus female DNA, we have observed loss of sex chromosome separation if the product of a single amplification is used. Although single products might be adequate for some assays, following problems with our CGH studies, we have since generated two products per sample so cannot comment on the reliability of a single product.

Acknowledgements

We would like to thank Suzie Little, Alan McIntyre, and Catherine Jones for their critical comments on the manuscript.

5. REFERENCES

1. Bates AS, Farrell WE, Bicknell EJ, *et al.* (1997) *J. Clin. Endocrinol. Metab.* **82**, 818–824.
2. Thakker RV, Pook MA, Wooding C, Boscaro M, Scanarini M & Clayton RN (1993) *J. Clin. Invest.* **91**, 2815–2821.
3. Simpson DJ, Bicknell EJ, Buch HN, Cutty SJ, Clayton RN & Farrell WE (2003) *Genes Chromosomes Cancer*, **37**, 225–236.
★ 4. Dean FB, Hosono S, Fang L, *et al.* (2002) *Proc. Natl. Acad. Sci. U. S. A.* **99**, 5261–5266. – *First description of WGA using multiple displacement amplification.*
5. Klein CA, Schmidt-Kittler O, Schardt JA, Pantel K, Speicher MR & Riethmuller G (1999) *Proc. Natl. Acad. Sci. U. S. A.* **96**, 4494–4499.
6. Nelson DL, Ledbetter SA, Corbo L, *et al.* (1989) *Proc. Natl. Acad. Sci. U. S. A.* **86**, 6686–6690.
7. Ludecke HJ, Senger G, Claussen U & Horsthemke B (1989) *Nature*, **338**, 348–350.
★ 8. Telenius H, Carter NP, Bebb CE, Nordenskjold M, Ponder BA & Tunnacliffe A (1992) *Genomics*, **13**, 718–725. – *First paper describing degenerate-oligonucleotide-primed PCR.*
★ 9. Zhang L, Cui X, Schmitt K, Hubert R, Navidi W & Arnheim N (1992) *Proc. Natl. Acad. Sci. U. S. A.* **89**, 5847–5851. – *First paper describing primer-extension pre-amplification.*

10. Lucito R, Nakimura M, West JA, *et al.* (1998) *Proc. Natl. Acad. Sci. U. S. A.* **95**, 4487–4492.
11. Liu CL, Schreiber SL & Bernstein BE (2003) *BMC Genomics*, **4**, 19.
12. Dietmaier W, Hartmann A, Wallinger S, *et al.* (1999) *Am. J. Pathol.* **154**, 83–95.
13. Kittler R, Stoneking M & Kayser M (2002) *Anal. Biochem.* **300**, 237–244.
14. Tanabe C, Aoyagi K, Sakiyama T, *et al.* (2003) *Genes Chromosomes Cancer*, **38**, 168–176.
★★★ 15. Hughes S, Arneson N, Done S & Squire J (2005) *Prog. Biophys. Mol. Biol.* **88**, 173–189. – *A good review of the current methods available for WGA.*
16. Cardoso J, Molenaar L, de Menezes RX, *et al.* (2004) *Nucleic Acids Res.* **32**, e146.
★ 17. Hosono S, Faruqi AF, Dean FB, *et al.* (2003) *Genome Res.* **13**, 954–964. – *Application of multiple displacement amplification to clinical samples.*
18. Hughes S, Lim G, Beheshti B, *et al.* (2004) *Cytogenet. Genome Res.* **105**, 18–24.
19. Lage JM, Leamon JH, Pejovic T, *et al.* (2003) *Genome Res.* **13**, 294–307.
20. Luthra R & Medeiros LJ (2004) *J. Mol. Diagn.* **6**, 236–242.
21. Paez JG, Lin M, Beroukhim R, *et al.* (2004) *Nucleic Acids Res.* **32**, e71.
22. Hellyer TJ & Nadeau JG (2004) *Expert Rev. Mol. Diagn.* **4**, 251–261.
23. Saunders RD, Glover DM, Ashburner M, *et al.* (1989) *Nucleic Acids Res.* **17**, 9027–9037.
24. Barker DL, Hansen MS, Faruqi AF, *et al.* (2004) *Genome Res.* **14**, 901–907.
25. Gribble S, Ng BL, Prigmore E, Burford DC & Carter NP (2004) *Chromosome Res.* **12**, 143–151.
26. Langmore JP (2002) *Pharmacogenomics*, **3**, 557–560.
★ 27. Devries S, Nyante S, Korkola J, *et al.* (2005) *J. Mol. Diagn.* **7**, 65–71. – *Use of WGA to study paraffin-embedded tissue.*
28. Douglas EJ, Fiegler H, Rowan A, *et al.* (2004) *Cancer Res.* **64**, 4817–4825.
29. Beheshti B, Braude I, Marrano P, Thorner P, Zielenska M & Squire JA (2003) *Neoplasia* **5**, 53–62.
30. Zielenska M, Marrano P, Thorner P, *et al.* (2004) *Cytogenet. Genome Res.* **107**, 77–82.
31. Wang G, Maher E, Brennan C, *et al.* (2004) *Genome Res.* **14**, 2357–2366.
32. Sellick GS, Longman C, Tolmie J, *et al.* (2004) *Nucleic Acids Res.* **32**, e164.

CHAPTER 7

DNA linear amplification

Chih Long Liu, Bradley E. Bernstein and Stuart L. Schreiber

Department of Chemistry and Chemical Biology, Harvard University, 12 Oxford St., Cambridge, Massachusetts 02138, USA

1. INTRODUCTION

Amplification of nucleic acids has become a mainstay of molecular biology. It permits genomic and transcriptional analysis when the amount of tissue or number of cells being studied becomes limiting, and is invaluable in studies where the biology of the specimens being investigated severely limits the amount of nucleic acids available. It is also particularly important for large-scale, high-throughput genomic studies, since these studies typically use microarrays or other high-throughput assays that often require microgram amounts of nucleic acids. Furthermore, these studies may employ a large matrix of many different conditions and/or time points, which may make it prohibitively expensive to generate sufficient amounts of unamplified material.

One example of such a genomics application is the ChIP–chip method, where DNA recovered from chromatin immunoprecipitation (ChIP) of cell lysate is used for subsequent analysis on DNA microarrays. This method (for review, see 1) is typically used to identify transcription factor binding sites and to map histone variants, histone post-translational modification patterns, or any other interesting epitope within the genome. *Fig. 1* shows an example of the ChIP–chip method using spotted microarrays.

ChIP, however, typically yields DNA in the nanogram range, which is insufficient for most DNA microarray applications. Laboratories using the ChIP–chip technique typically employ an exponential amplification method, such as ligation-mediated PCR (LM–PCR) (2) or random PCR (R–PCR) (3–5), to obtain the quantities necessary for microarray analysis. R–PCR involves the annealing of primer adaptors with a 5′ conserved end and a 3′ degenerate end to the template DNA, followed by extension and subsequent PCR with primers complementary to the 5′ conserved ends. These exponential amplification methods adequately fulfill the needs of ChIP–chip analysis and are still frequently used, particularly when amplification from subnanogram amounts are

Whole Genome Amplification: Methods Express (S. Hughes and R. Lasken, eds.)
© Scion Publishing Limited, 2005

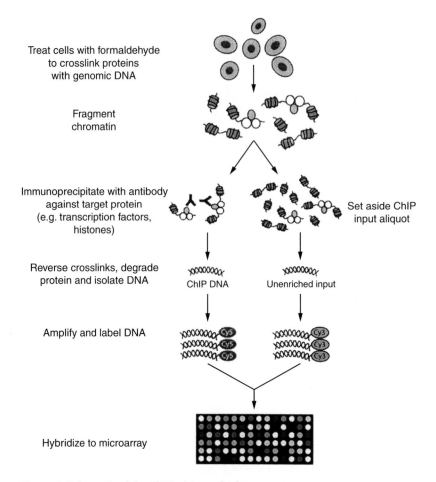

Treat cells with formaldehyde
to crosslink proteins
with genomic DNA

Fragment
chromatin

Immunoprecipitate with antibody
against target protein
(e.g. transcription factors,
histones)

Set aside ChIP
input aliquot

Reverse crosslinks, degrade
protein and isolate DNA

ChIP DNA   Unenriched input

Amplify and label DNA

Hybridize to microarray

Figure 1. Schematic of the ChIP–chip method.
Cells are first crosslinked with formaldehyde prior to lysis and DNA fragmentation.
Following fragmentation, chromatin is then incubated with antibodies and subsequently
immunoprecipitated with Protein A or Protein G beads, which bind to the F_c segment of
the antibodies. Following elution from the beads, reversal of crosslinks, and proteinase K
digestion, ChIP samples are typically phenol-chloroform extracted, ethanol precipitated,
and then treated with RNase A to eliminate RNA that has been carried over from the
immunoprecipitation. The ChIP samples and the unenriched input material are then
amplified and labeled with fluorescent dyes. The ChIP sample is subsequently hybridized,
along with the unenriched input, on a spotted microarray.

desired. However, they have a number of shortcomings, in particular with regard
to amplification fidelity and to a lesser extent with dynamic range compression,
and with their inefficient amplification of short nucleic acids such as those less
than 250 bp in length (6).

2. METHODS AND APPROACHES

2.1. DNA linear amplification

The DNA linear amplification method described here has been designed to address the shortcomings of exponential amplification. This method uses a linear amplification approach, based on the *in vitro* transcription (IVT) of template DNA by RNA polymerase from the T7 phage, a common strategy employed in a number of published RNA amplification protocols (7–9; for review, see 10). This linear amplification approach successfully addresses the amplification fidelity issues raised with an exponential amplification approach (6). The amplification method presented in this chapter was designed primarily to address shortcomings of the R-PCR protocol with the ChIP–chip method. Thus, optimizations made to this method have been done with this particular application in mind. For other applications that require DNA to be the end point, reverse transcription is a necessary step that increases the cost and complexity of necessary molecular biological manipulations to the sample, when compared with PCR. However, IVT amplification does offer improved fidelity and a much higher maximum yield per single reaction (6). Thus, other techniques that require amplification of complex mixtures of randomly fragmented genomic DNA can also benefit from this method.

2.2. General strategy

This section describes in detail the general strategy employed by this linear amplification method. This method takes nanogram quantities of dsDNA as the starting material and generates microgram amounts of amplified RNA. An overall schematic diagram is shown in *Fig. 2*. Briefly, the strategy is to add a 3′ conserved end to the template dsDNA, using terminal deoxynucleotidyl transferase (TdT) tailing, which permits the addition of a T7 promoter sequence in the subsequent second-strand synthesis step. IVT can then utilize this newly appended T7 promoter and linearly amplify the template dsDNA, producing amplified RNA product.

2.3. Considerations for the starting dsDNA template

The ideal starting material for this method is dsDNA template in the 100–1000 bp size range with 3′ protruding or 3′ blunt ends. The 3′ end must have a free hydroxyl (OH) group, since TdT does not add residues to template strands with 3′ phosphate groups. Some restriction digests and other DNA fragmentation methods such as sonication and nuclease digestion may leave behind 3′ phosphate groups on a significant proportion of the DNA molecules. These phosphate groups need to be removed prior to the tailing reaction; failure to do so may result in poor amplification.

One other consideration about 3′ ends is their efficacy in serving as an efficient template for TdT-mediated polynucleotide tailing. The efficiency of the

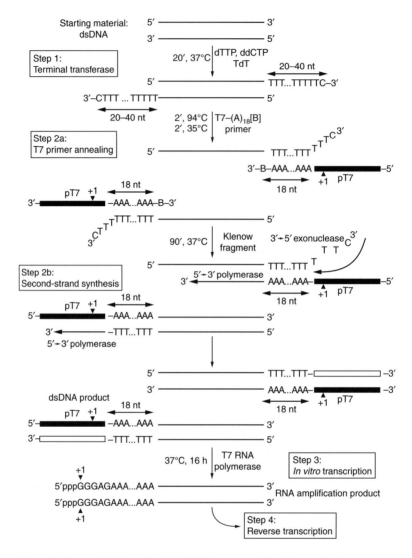

Figure 2. General strategy for the DNA linear amplification method.
Starting with dsDNA template, TdT is used to add a poly(dT) tail to the 3′ ends of the template. This tail subsequently provides a conserved binding site for the annealing of T7 promoter (pT7)–poly(dA) primer adaptors. Following subsequent second-strand synthesis using the large fragment of DNA polymerase I (Klenow fragment), one pair of dsDNA templates, with each pair member representing one of the two complementary strands of the dsDNA, is generated, with a T7 promoter at the 5′ end of the amplicon. In the subsequent IVT step, RNA is transcribed from this template in an isothermal reaction, producing an RNA amplification product consisting of both strands of the original dsDNA template in high microgram quantities. Note that each RNA strand will contain a short sequence from the T7 promoter and a poly(A) tract, 5′ relative to the amplicon. Reprinted with permission from (6).

TdT enzyme is maximal on 3′ protruding ends and good on 3′ blunt ends. However, 3′ recessed ends are only 'very reluctantly' tailed, and a mixture of templates containing all three types of 3′ end will likely result in tailed products enriched with previously 3′ protruding ends and depleted of previously 3′ recessed ends (11). The end result is a larger spread in the size distribution of poly(dT) tail lengths, which may negatively impact subsequent amplification fidelity if the relationship between the template sequence and the nature of the 3′ end is nonrandom. It is therefore recommended that, in applications that are sensitive to this issue, templates containing a significant proportion of 3′ recessed ends are filled in with Klenow fragment DNA polymerase prior to the TdT tailing step.

The size range of approximately 100–1000 bp is considered to be optimal as this is the size range that was extensively tested with this method when it was developed. Amplification of fragments smaller than 100 bp is limited by the column-based purification steps, listed in *Protocols 5* and *6*, which rely on the Qiagen RNeasy columns and MinElute columns. The MinElute columns have a lower limit of 70 bp, while the RNeasy columns have a tested limit of 100 bp (which is lower than the 200 bp manufacturer-specified limit). While *Protocols 5* and *6* were developed with Qiagen columns, alternative reaction cleanup columns or other methods can be substituted.

Amplification of material smaller than 100 bp will also be less efficient in terms of mass. This is because the rate-limiting step in IVT is initiation of transcription (12), and because there is a greater molar ratio of template molecules to T7 RNA polymerase compared with an equivalent mass of a higher-molecular-weight template. Here, the low-molecular-weight template would require more initiation of transcription events per mass unit compared with a high-molecular-weight template, and the usual result is a significantly lower mass yield for the low-molecular-weight template (see Section 3.1 for details).

Amplification of fragments larger than 1000 bp has not been extensively tested. This should not be a particular problem for templates ranging up to 4–5 kb in size, since T7 RNA polymerase is highly processive (12). Extremely large fragments (>5 kb) are perhaps best amplified by other techniques (e.g. strand displacement, 13).

2.4. Using this method for ChIP–chip experiments

Since this method was originally designed to address the amplification needs of ChIP–chip experiments, this section will discuss in detail the considerations necessary to optimize the amplification fidelity and yield of the starting material. However, this approach will be appropriate for many applications that require several micrograms of DNA from nanogram quantities of starting DNA.

ChIPs frequently yield low amounts of DNA, which may be difficult to quantify via usual methods, such as absorbance at 260 nm with a spectrophotometer. Most commonly used spectrophotometers have a lower reliable detection limit of 0.01 absorbance units and are limited to a minimum cuvette volume of 50–100 μl, translating to 0.5 ng/μl for dsDNA. Consequently, it becomes quite difficult to detect and quantify reliably the presence of nucleic acid when the total yield is less than 250 ng, without sacrificing an excessive proportion of the total sample in the measurement. Reliable measurement of a ChIP mass yield is important for maximizing amplification yield for this method. Therefore, we recommend using a more sensitive instrument or a spectrophotometer designed to handle cuvettes with small volumes (<10 µl). Since these instruments require five- to tenfold less sample per measurement, the lower limit then becomes 25–50 ng of ChIP yield for reliable measurement. For extremely low ChIP yields (<25 ng), quantification via fluorescence (e.g. using PicoGreen (Molecular Probes) with the Turner BioSystems TD-700 fluorometer) is the recommended method and should reliably measure samples down to a lower limit of 2.5 ng starting material.

There are a couple of additional technical considerations when this method is used with the ChIP–chip method. Calf intestinal alkaline phosphatase (CIP) treatment (*Protocol 1*) is presented as an optional protocol, but is necessary if the ChIP–chip method employs fragmentation methods (e.g. sonication or nuclease digestion) that leave behind a significant proportion of 3′ phosphate groups in the template DNA. This is discussed in further detail in section 4.1.3. Additionally, since *Protocol 1* is typically carried out immediately after RNase A digestion in the ChIP–chip method, users may be concerned about sufficient elimination of RNase A for the remainder of *Protocol 1* and in *Protocols 2* and *3*, in time for the IVT step (*Protocol 4*). The three MinElute cleanup steps (*Protocol 6*, performed at the end of *Protocols 1, 2*, and *3*), are usually sufficient to remove enough residual RNase A for it not to pose a problem. Furthermore, the Ambion IVT enzyme mix used in *Protocol 4* contains an RNase inhibitor. Care should be taken, however, to ensure that barrier tips are used, and ideally separate pipettes should be used when handling the RNase A enzyme to avoid RNase contamination. Troubleshooting RNase contamination is discussed in section 4.1.1.

2.5. Controls for new users

To maximize the success of this method, two positive controls are strongly suggested for new users. The troubleshooting section will discuss in greater detail a number of methods that can be used to resolve low-yield problems.

2.5.1. *Protocols 2 and 3*: positive amplification control

This control enables the user to distinguish sample-specific problems from protocol-implementation issues. Here, the starting material is 50 ng of blunt-ended PCR product in the 100–1000 bp range (preferably around 200–500 bp). If this control is used as part of a ChIP–chip experiment, the PCR product can be subject to RNase A and CIP treatment for troubleshooting purposes, although in practice these treatments are usually not necessary. Yields should typically range from 30 to 60 µg (with a maximum of 80–100 µg observed), depending on the size of the PCR product, protocol implementation, and quality of the reagents used (particularly the nucleotides used for IVT (in *Protocol 4*), which are highly sensitive to freeze–thaw cycles).

2.5.2. *Protocol 4*: positive IVT control

This control enables the user to distinguish sample-specific problems from IVT-specific problems (such as RNase contamination). Here, the IVT starting material is 250 ng of the pTRI-Xef linearized plasmid provided with the Ambion IVT kit (see *Protocol 4*). If not using the kit, an appropriate amount of a dsDNA template that contains the pT7 promoter, with prior history of use as a successful T7 RNA polymerase template, should be used. Yields should typically range from 100 to 140 µg, limited by exhaustion of the nucleotides in the reaction mixture and the

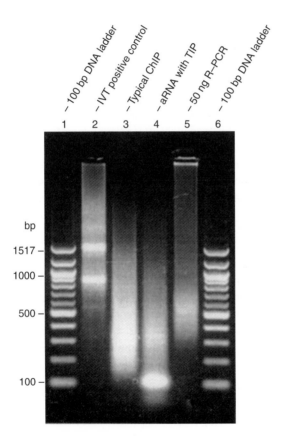

Figure 3. Amplification products on a non-denaturing 2% agarose gel in TAE buffer, stained with ethidium bromide.
A 100 bp ladder (500 ng, New England Biolabs) or 650 ng of samples were loaded. (*Lanes 1 and 6*) 100 bp ladder; (*lanes 2–4*) antisense RNA (aRNA) amplified using the method presented in this chapter, with the following templates: (*lane 2*) IVT positive control (pTRI-Xef) (see *Protocol 4*); (*lane 3*) a typical ChIP sample (*Saccharomyces cerevisiae* bis-acetyl histone H3 on lysine 9 and 14) from sonicated genomic DNA; (*lane 4*) amplification from 5 ng of a genomic digest of *S. cerevisiae* with *Rsa*I, showing a strong band of template-independent product (TIP) of approximately 100 bp; (*lane 5*) the R–PCR product, for comparison, from a 50 ng amplification of the same template used to generate aRNA in *lane 4*. The most remarkable attribute of *lane 5* is the near absence of DNA smaller than 250 bp.

rated 100 µg binding capacity of the Qiagen RNeasy column. Any yield somewhat less than this may suggest RNase contamination, evaporation of the reaction mixture during the long incubation period (due to the small reaction volume), or poor-quality reagents (particularly the nucleotides). Visualization on a non-denaturing agarose gel should yield two intense bands of approximately 0.9 and 1.5 kb, as shown in *Fig. 3* (*lane 2*).

3. RECOMMENDED PROTOCOLS

All protocols presented in this section are typically performed in the order presented when used in conjunction with the ChIP–chip method. *Protocol 1*, CIP treatment, is an optional but strongly recommended protocol for removing 3′ phosphate groups, since most genomic DNA fragmentation methods (i.e. sonication, micrococcal nuclease digestion, and certain restriction digests) will produce a significant proportion of 3′ phosphate groups within the mixture of fragmented genomic DNA. This protocol is compatible with the presence of RNase A and can be carried out immediately after RNase A digestion of RNA carried over from the ChIP, without any intermediate cleanup step.

The main linear amplification protocol begins with dsDNA and produces an RNA amplification product, as shown in *Fig. 2*. The procedure can be broken down into three main stages:

(i)   the tailing reaction with terminal transferase;
(ii) second-strand synthesis with Klenow fragment polymerase; and
(iii) IVT.

The tailing reaction involves the addition of a short (20–40 nt) poly(dT) tail to the template DNA. This poly(dT) tail provides a conserved 3′ element that permits annealing of a T7 primer adaptor during second-strand synthesis, as described in *Protocol 3*. The inclusion of the dideoxynucleotide acts as a tail terminator in the reaction mixture and is necessary to maintain a tight size distribution of the poly(dT) tail.

Protocol 1

CIP treatment of samples with terminal 3′ phosphate groups

Equipment and Reagents

- CIP (2.5 units/µl) (New England Biolabs)
- 10× NEB Buffer 3 (100 mM NaCl; 50 mM Tris-HCl (pH 7.9); 10 mM MgCl$_2$; 1 mM dithiothreitol (DTT)) (New England Biolabs). This is typically supplied with the CIP enzyme.
- TE (10 mM Tris-HCl (pH 8.0); 1 mM EDTA)
- Nuclease-free water (Ambion)
- Water bath or heat block set to 37°C

Method

1. Prepare the CIP reaction mixture by combining, for every 10 µl[a]:
   1 µl of 10× NEB Buffer 3 (1× final concentration)
   template DNA to a maximum of 50 ng/µl final concentration[b,c,d]
   0.25 µl of CIP
   nuclease-free water up to a final volume of 10 µl

2. Incubate the reaction at 37°C for 1 h[e].

3. Clean up the reaction with the Qiagen MinElute Kit (follow *Protocol 6*). Be sure to elute in 20 µl, which is twice the manufacturer's suggested elution volume.

Notes

[a]Each reaction can be scaled up to 100 µl per tube. A typical reaction volume is 30–40 µl.

[b]Template dsDNA samples should be suspended in water, TE, or 1 × NEB Buffer 3 (which will not interfere with RNase digestion). If suspended in NEB Buffer 3, addition of extra NEB Buffer 3 is unnecessary. Template dsDNA samples must not be denatured.

[c]There should be no more than 500 ng of template dsDNA per 10 µl reaction volume.

[d]Samples processed from the ChIP–chip method are typically quantified by one of several methods mentioned in section 2.4. The exact concentration of template dsDNA is not very critical for *Protocol 1*. However, it is extremely important that the concentration of the template DNA be determined accurately for *Protocol 2*.

[e]Some users might consider combining CIP treatment and RNase digestion to save time and labor, when incorporating the ChIP–chip method with DNA linear amplification. This can be done, but may diminish yield, depending on the amount of undigested RNA present. If the amount of undigested RNA is large (see *Fig. 1*), the yield may drop by up to 50%. This is likely due to the undigested RNA competing with template DNA for CIP activity, resulting in less-efficient removal of the 3′ phosphate groups. However, ChIP samples are unlikely to contain large amounts of carry-over RNA and are less likely to be affected by this issue.

Protocol 2

Tailing reaction with terminal transferase

Equipment and Reagents
- TdT[a] (20 units/μl) (New England Biolabs)
- 5× TdT buffer[b] (1 M potassium cacodylate[c]; 125 mM Tris-HCl (pH 6.6); 1.25 mg/ml bovine serum albumin) (Roche)
- 8% dideoxynucleotide tailing solution[d] (92 μM dTTP; 8 μM ddCTP (Invitrogen))
- 5 mM cobalt chloride (Roche)
- 0.5 M EDTA (pH 8.0)
- Mineral oil (molecular biology grade)
- Water bath or heat block set to 37°C

Method
1. Prepare the TdT reaction mixture by combining:

   2 μl of 5× TdT buffer (1× final concentration)

   0.5 μl of 8% dideoxynucleotide tailing solution

   1.5 μl of 5 mM cobalt chloride (0.75 mM final concentration)

   5 μl of template DNA[e] (maximum 7.5 ng/μl final concentration)

   1 μl of TdT enzyme (2 units/μl final concentration). TdT should be the *last* reagent added to the mixture

2. Add one to two drops of mineral oil to the top of the mixture to prevent evaporation during incubation.

3. Incubate the reaction at 37°C for 20 min.

4. Stop the reaction by adding 2 μl (per 10 μl reaction volume) of 0.5 M EDTA (pH 8.0).

5. Clean up the reaction using the Qiagen MinElute Kit (follow *Protocol 6*). It is preferable to minimize the amount of mineral oil carried over from the reaction, although trace amounts are acceptable. When using a 10 μl reaction volume, add 10 μl of water to bring the volume up to 20 μl prior to following *Protocol 6*[f].

Notes

[a]The New England Biolabs (NEB) recombinant enzyme is the preferred enzyme source. Use a recombinant enzyme, since an enzyme derived from a natural source (e.g. Roche, typically derived from calf thymus) may have lot-dependent variation and may result in unpredictable or lowered IVT yields. Furthermore, the NEB recombinant enzyme is preferred over the Roche recombinant enzyme, because the latter typically yields 50% less amplification product with the volumes specified in *Protocol 2*.

[b]Note that this is *not* the buffer supplied with the NEB enzyme. This is actually the buffer supplied with either of the Roche enzymes (see [a]). The NEB TdT enzyme comes supplied with NEB Buffer 4 (50 mM potassium acetate; 20 mM Tris-acetate; 10 mM magnesium acetate; 1 mM DTT) and 2.5 mM cobalt chloride. The DTT in this buffer will precipitate the cobalt chloride and inhibit the reaction.

[c]Cacodylate is a methylated form of arsenic. While this form is less toxic than other forms of arsenic, it should still be treated as a toxic reagent and handled accordingly. Employ waste disposal practices appropriate for your institution.

[d]Avoid subjecting the 8% tailing solution to more than three freeze–thaw cycles, as we have found that additional freeze–thaw cycles will further degrade the nucleotides and reduce the efficiency of the reaction.

[e]The maximum amount of template DNA is approximately 1 pmol of template molecules. This corresponds to an approximate maximum of 75 ng for a mixture of template DNA with an average size range of 250 bp. The tested range is 2.5–75 ng of DNA per 10 µl reaction volume. Scale up the reaction volume accordingly for higher starting amounts, typically to 20 µl. For ChIP samples, an accurate concentration is critical – underestimation of the concentration will result in lowering the yield to as little as 5–10% of the amount that is typical for *Protocol 2*, because much of the template DNA will not be tailed. If you are not sure how much you have and cannot accurately quantify your samples, scale up to a 20 µl volume.

[f]Be sure to elute in 20 µl, which is twice the manufacturer's suggested elution volume of 10 µl, to achieve sufficient yields. Eluting in 10 µl may reduce the yield by up to 50%, possibly because the MinElute columns have a decreased recovery yield for nanogram quantities of DNA.

Second-strand synthesis involves synthesis of the second strand of the template DNA. At this stage, the strand-displacement activity of the Klenow fragment polymerase separates the two strands of the template DNA, after which the enzyme performs fill-in 5′→3′ polymerization. Its 3′→5′ exonuclease activity may also remove the 3′ overhanging poly(dT) tails (see *Fig. 2*), although the efficiency of this activity will vary based on the length of the poly(dT) tail. If template-independent product is generated, as mentioned in section 4.1.4 and shown in *Fig. 3* (*lane 4*), refer to *Table 1* for adjustments to the reaction volumes of this protocol.

The IVT step is the stage at which linear amplification is performed. Because the T7-based IVT proceeds as an isothermal reaction, it linearly amplifies the template DNA, producing antisense RNA (aRNA), i.e. each strand of RNA produced is antisense to the original template strand. Since both strands are amplified, this distinction is usually not important and is affected only by the location of the T7 promoter and poly(A) tract on the aRNA. Note that RNA is produced at the end of this protocol, so practice correct techniques to maintain an RNase-free environment.

Table 1. Second-strand synthesis with limiting primer amounts

DNA (ng)	T7 primer (µl)[a]	NEB Buffer 2 (µl)	5 mM dNTPs (µl)	Water (µl)	Tailed DNA (µl)	Klenow (µl)	Total volume (µl)
>75	0.60 (25 µM)	5.0	2.0	20.40	20.0	2.0	50
50–75	0.30 (25 µM)	2.5	1.0	0.20	20.0	1.0	25
25	0.15 (25 µM)	2.5	1.0	0.35	20.0	1.0	25
10[b]	1.50 (1 µM)	1.0	0.4	0.20	6.5	0.4	10
5[b]	0.75 (1 µM)	1.0	0.4	0.95	6.5	0.4	10
2.5[b]	0.38 (1 µM)	1.0	0.4	1.32	6.5	0.4	10

Using limiting amounts of primer is highly advisable when amplifying from very small amounts of starting material. Not only will this decrease the amount of primer-dimer product, but it may also increase the yield of the desired amplification product. The table gives the single reaction volumes to use for a suggested mass range of starting material.

[a]Spin down the tubes every 30 min during the 37°C incubation step if using a thermal cycler that does not have a heated lid.

[b]The tailed DNA will have to be dried down in a vacuum centrifuge to the volume indicated for reaction volumes that total 10 µl.

Protocol 3

Second-strand synthesis with Klenow fragment polymerase

Equipment and Reagents
- DNA polymerase I Klenow fragment (5000 units/ml) (New England Biolabs)
- NEB Buffer 2 (50 mM NaCl; 10 mM Tris-HCl (pH 7.9); 10 mM $MgCl_2$; 1 mM DTT) (New England Biolabs). This is typically supplied with the Klenow fragment enzyme.
- 25 µM T7-A_{18}B primer
  (5′-GCATTAGCGGCCGCGAAATTAATACGACTCACTATAGGGAG(A)$_{18}$[B]-3′)[a]
- 5.0 mM dNTP mix[b] (Invitrogen)
- 0.5 M EDTA (pH 8.0)
- Nuclease-free water (Ambion)
- Thermal cycler

Method
1. Prepare the second-strand reaction mixture in thermal-cycler-compatible tubes by combining[c]:

   2.5 µl of 10× NEB Buffer 2[d] (1× final concentration)
   1 µl of 5 mM dNTP solution (200 µM final concentration)
   0.3 µl of 25 mM T7-A_{18}B primer (300 nM final concentration)
   20 µl of poly(dT)-tailed template DNA (elution volume from *Protocol 6*)
   nuclease-free water to a final volume of 24 µl (0.2 µl)

   The volumes correspond to a typical reaction volume of 25 µl, taking into account the 1 µl of Klenow polymerase to be added in step 2. Do *not* use mineral oil. Trace amounts of mineral oil may interfere with the IVT reaction (see *Protocol 4*).

2. Incubate the reaction in the thermal cycler using the following program: 94°C for 2 min to denature; ramp –1°C/s to 35°C; hold at 35°C for 2 min to anneal primers; ramp –0.5°C/s to 25°C; hold for 45 s (pausing longer, up to 5 min, is permitted if the time is needed for adding Klenow to a large number of samples). Remove the tubes, add the amount of Klenow DNA polymerase indicated in *Table 1*, spin the tubes briefly, return them to the cycler, and incubate at 37°C for 90 min to fill in the second strand.

3. Stop the reaction by adding 0.5 M EDTA (pH 8.0) to 50 mM final concentration (2.5 µl for a 25 µl reaction volume).

4. Clean up the reaction with the Qiagen MinElute Kit (follow *Protocol 6*). Be sure to elute in 20 µl, which is twice the manufacturer's suggested elution volume.

Notes
[a][B] stands for any base other than A. The primer thus consists of a mix of primers that end in C, G or T. This primer adaptor should be obtained by high-pressure liquid chromatography, polyacrylamide gel electrophoresis, or an equivalent purification method.

[b]This is a deoxynucleotide mixture containing 5 mM each of dATP, dCTP, dTTP, and dUTP. We have found that additional freeze–thaw cycles will further degrade the dNTPs and reduce the reaction yield.

[c]The volume should be scaled up to 50 µl if *Protocol 2* was scaled up to 20 µl. The volumes correspond to a typical reaction volume of 25 µl. Refer to *Table 1* for any necessary adjustments.

[d]In early 2004, New England Biolabs switched the supplied buffer for the Klenow enzyme from EcoPol Buffer (10 mM Tris-HCl (pH 7.5); 5 mM magnesium chloride; 7.5 mM DTT) to NEB Buffer 2. The NEB Buffer 2 performs at least equivalently, if not better, than the EcoPol Buffer, which had to be pre-warmed to 37°C to dissolve any precipitated DTT.

Protocol 4

IVT

Equipment and Reagents
- T7 Megascript Kit (containing 75 mM each of ATP, CTP, GTP, and UTP nucleotide solutions[a]; pTRI-Xef (0.5 mg/ml) control template (optional); nuclease-free water; 10× reaction buffer[b]; 10× enzyme mix (T7 RNA polymerase and a proprietary RNase inhibitor)) (Ambion)
- Air incubator set to 37°C

Method
1. If continuing from *Protocol 3*, dry down the eluate from 20 to 8 µl in a vacuum centrifuge at medium heat for 10–12 min.

2. Prepare the IVT reaction mixture in 0.2 ml RNase-free tubes by combining:
   8 µl of 75 mM NTP mix[a]
   2 µl of 10× reaction buffer[b]
   2 µl of enzyme mix[c]
   8 µl of eluate from *Protocol 3*

3. Incubate at 37°C overnight (acceptable range is 5–20 h) in the air incubator or in a thermal cycler with a heated lid. Tubes of 0.2 ml are used to minimize vapor volume.

Notes

[a]If using a new kit, combine the NTP solutions into one tube, then aliquot back out into the four tubes. In the first three freeze–thaw cycles, yields drop approximately 10–15% after each cycle. If the NTPs go through more than three freeze–thaw cycles, we have found that each subsequent freeze–thaw cycle may additionally drop the yield by as much as 50%.

[b]The reaction buffer should be warmed to room temperature first. Adding cold buffer to the template DNA may risk precipitation of the DNA.

[c]If your template DNA is small (<300 bp), you can try boosting the reaction by increasing the enzyme mix to 2.4 µl and decreasing the NTP mix to 7.6 µl. The reaction yield may increase by 10–30% due to the more favorable stoichiometric ratio of enzyme to template DNA in the boosted reaction. Note, however, that this may lower your maximum theoretical yield, so this step is not recommended for larger DNA templates.

After the IVT reaction is complete, the aRNA product is cleaned up using the Qiagen RNeasy Kit. The following protocol is based on the manufacturer's protocol for cleaning up RNA reactions, with minor modifications.

If the aRNA is to be used for subsequent microarray experiments, the reverse transcription reaction should be primed with 5 µg pd(N)$_6$ random hexamer primers (Amersham Biosciences). The presence of oligo(dT) primers will not interfere with reverse transcription, but oligo(dT) primers will not prime the reaction correctly.

The Qiagen MinElute protocol is based on the manufacturer's protocol, except that the elution volume has been doubled to 20 µl, due to the small amounts of DNA being purified at each step. Without this increase in elution volume, yields may drop by as much as 50%.

Protocol 5

Sample purification

Equipment and Reagents
- RNeasy Mini Kit (containing RNeasy columns; Buffer RLT; Buffer RPE (with 95 or 100% RNase-free ethanol added)[a]; RNase-free water; 2 ml RNase-free collection tubes; 1.5 ml RNase-free collection tubes) (Qiagen)
- 14.2 M β-Mercaptoethanol (Sigma)
- 100% RNase-free ethanol (Sigma)

Method
1. Prepare the Buffer RLT master mix[b] by combining the following in a 1.5 ml RNase-free microcentrifuge tube for each IVT reaction:
     350 µl of Buffer RLT
     80 µl of RNase-free water
     3.5 µl of β-mercaptoethanol

2. Transfer the contents of the IVT reaction to the 1.5 ml tube and vortex briefly. Low-retention, aerosol-barrier, RNase-free pipette tips are highly recommended here, since the RNA concentration in the IVT reaction tubes may be as high as 5 µg/µl.

3. Add 250 µl of ethanol (95–100%) and mix well by pipetting. Do not spin the tubes down.

4. Apply the sample (~700 µl) to an RNeasy mini spin column sitting in a collection tube. Centrifuge for 15 s at 8000 g. Discard flow-through.

5. Transfer the RNeasy column to a new 2 ml collection tube. Add 500 µl of Buffer RPE (which must contain ethanol) and centrifuge for 15 s at 8000 g. Discard flow-through but re-use collection tube.

6. Pipette 500 µl of Buffer RPE into the RNeasy column and centrifuge for 1 min at maximum speed in a microcentrifuge.

7. Remove flow-through and pipette another 500 µl of Buffer RPE on to the column. Centrifuge for 2 min at maximum speed in a microcentrifuge to dry the column completely[c].

8. Transfer the RNeasy column into a new 1.5 ml RNase-free collection tube, taking care not to carry over any flow-through from the 2 ml collection tube.

9. Add 30 µl of RNase-free water directly on to the membrane of the RNeasy column. Centrifuge for 1 min at 8000 g to elute. Repeat this step if the expected yield is ≥30 µg.

10. Check RNA concentration and quality by measuring the absorbance at 260 nm and 260/280 nm, and by running a sample on a 1–2% agarose gel. A denaturing gel should only be used if the size distribution of the aRNA needs to be determined accurately. *Fig. 3 (lane 4)* indicates a typical 2% gel result.

Notes

[a]This protocol consumes 50% more Buffer RPE than the manufacturer's standard protocol. It may be necessary to order additional Buffer RPE separately for a sufficient supply. If necessary, 80% RNase-free ethanol may be substituted, although this has not been extensively tested.

[b]This mix can be prepared up to a week in advance and aliquoted into 1.5 ml microcentrifuge tubes.

[c]This is an additional wash that is not in the Qiagen protocol. If the RNA is to be used for microarray work or other applications involving fluorescence, we have found this additional wash to be necessary to remove remaining trace amounts of guanidine thiocyanate that would otherwise contaminate the eluted RNA and cause increased background noise in fluorescence applications such as microarrays.

Protocol 6

Qiagen MinElute Kit protocol[a]

Equipment and Reagents
- MinElute Kit (containing MinElute columns; Buffer ERC; Buffer PE (with 95% or 100% ethanol added; Buffer EB) (Qiagen)
- 3 M Sodium acetate (pH 5.0) (optional)

Method
1. In a sample of volume 20–100 µl, add 300 µl of Buffer ERC and mix thoroughly. If the sample is in less than 20 µl, make it up to 20 µl with nuclease-free water. If the sample is in more than 100 µl, split the sample into aliquots smaller than 100 µl and process each aliquot in its own column.

2. If the buffer color is orange or purple (i.e. pH>7.5), add 10 µl of 3 M sodium acetate (pH 5.0). If the buffer is yellow, as is typically the case for *Protocols 1* and *2*, no additional sodium acetate is needed.

3. Apply the sample mixture to the MinElute spin column (sitting in a 2 ml collection tube) and spin for 1 min at maximum speed in a microcentrifuge.

4. Discard the flow-through and add 750 µl of Buffer PE (which must contain ethanol). Spin for 1 min at maximum speed in a microcentrifuge.

5. Discard the flow-through and spin for 1 min at maximum speed in a microcentrifuge to dry the column.

6. Transfer the column to a fresh 1.5 ml nuclease-free microcentrifuge tube. Pipette 20 µl of Buffer EB directly on to the column membrane. Leave it to stand for at least 1 min and then spin for 1 min at maximum speed in the microcentrifuge to elute.

Notes
[a]This is based on the MinElute Reaction Cleanup Kit protocol provided in the MinElute Handbook supplied with the kit.

3.1. Expected yields

Depending on the starting amount and size distribution of the starting material, the typical yield is 5–90 µg, as detailed in *Table 2*. The amplification yield for a given starting amount is represented by a range, which is dependent on the quality and size distribution of the starting material. A more direct determinant of amplification yield will be based on the number of picomoles of template used. Initiation of the reaction is the rate-limiting step for TdT tailing (see *Protocol 2*) and IVT (see *Protocol 4*), so the number of template molecules present will more directly determine the actual amplification yield. For example, amplifying 50 ng of a 200 bp PCR product is likely to produce a lower mass yield than amplifying 50 ng

Table 2. Typical amplification yields

Input (ng)	Yield (µg)	Fold amplification
75	60–100	800–1333
50	50–90	1000–1800
25	30–70	1200–2800
10	20–50	2000–5000
5	10–25	2000–5000
2.5	5–10	2000–4000
50 (R–PCR)	15	300

of an 800 bp PCR product, even though the number of aRNA molecules produced may actually be greater for the 200 bp product than for the 800 bp product. For comparison, the R–PCR yield (30 cycles) of a 50 ng starting amount is also shown (see *Fig. 3, lane 5*, and *Table 2*).

3.2. Composition of the amplification product

Typically, a 1–2% agarose gel, containing or stained with ethidium bromide, is used to assess the composition and quality of the amplified RNA. An example of such a gel is shown in *Fig. 3*. Unless size distribution is crucial, it is usually not necessary for the gel to be denaturing. Within the resolution limits of the agarose gel, the amplified product may shift higher on the gel, in the order of 20–40 bp. This shift is to be expected due to the addition and tightness of the size distribution of the poly(A) tail, plus the 5 nt sequence added by the T7 promoter. The size distribution of the poly(A) tail becomes particularly evident in amplification products produced from a single-size template, such as PCR products or a restriction-digested plasmid. Occasionally, a substantial amount of a low-molecular-weight band may also appear near the bottom of the gel, at around 100 bp (see *Fig. 3, lane 4*). This band has been observed under certain amplification conditions, usually when the concentration of starting material is significantly less than that of the primer during second-strand synthesis (see *Protocol 3*). Although the composition of this band has not been tested, it likely represents the amplification product produced from IVT-valid template synthesized through the formation of primer dimers during second-strand synthesis (see *Protocol 3*). This side reaction consumes reagents and activity of the RNA polymerase enzyme, reducing the true effective yield of the desired amplification product. Therefore, *Table 1* has been included in *Protocol 3*, in which optimum primer-to-template mass ratios (roughly 5:1) have been established to minimize this side reaction.

The Qiagen MinElute Kit is used a number of times in this method to remove buffer salts, spent enzymes, free nucleotides, and any DNA products shorter than 40 nt. Material between 40 and 70 nt may also be removed, but at a lower efficiency.

3.3. Results

3.3.1. Summary of amplification fidelity

A detailed discussion of the validation procedure for amplification fidelity and reproducibility will not be repeated here, since it has been previously described in detail (6). However, key points from this discussion will be briefly mentioned. First, within the 100–700 bp range tested for this method, the size distribution of the amplification product is effectively preserved compared with R–PCR, which loses material below 250 bp and subsequently has reduced fidelity (see *Fig. 3, lane 5*). IVT amplification also produces an increased dynamic range that does not appear to impact negatively on its fidelity, improving the prospects of separating biological trends from the noise inherent in microarray experiments. This is another improvement over R–PCR, which often suffers from dynamic range compression. Furthermore, IVT amplification retains fidelity and representation of the starting material when compared with an unamplified Klenow-labeled standard.

3.3.2. Updates to this method

Since the initial publication of this method (6), a number of updates have been incorporated, as presented in this chapter. While most of the changes are minor, a number of more significant changes have been implemented. This method was optimized with the NEB Klenow enzyme for the second-strand synthesis step (see *Protocol 3*). Recently, NEB switched from supplying EcoPol buffer with this enzyme to Buffer 2. This switch has now been tested, with results indicating no change or a slight improvement (5–15%) in yields. Furthermore, the second-strand synthesis step now contains a detailed table (see *Table 1*) of optimized primer amounts and recommended final reaction volumes for a given starting amount. These measures combine to reduce the side reaction of low-molecular-weight material generated from speculated primer-dimer formation from excess primer (6, 7). Moreover, the default reaction volume has been reduced from 50 to 25 µl for the recommended starting amount of 50–75 ng, which permits the reduction of reagent costs without affecting yield. Finally, for the IVT step, amplification of low-molecular-weight material (i.e. <300 bp) has been optimized by boosting the amount of T7 enzyme used per reaction. The boost typically results in a 10–30% yield improvement, likely due to a more favorable stoichiometric ratio of the T7 enzyme to the DNA template.

3.4. ChIP–chip results

The method presented in this chapter has already been used in a number of published studies (14–17). An example of the results obtained from the ChIP–chip method, when used in conjunction with DNA linear amplification, is shown in *Fig. 4*. Here, the ChIP–chip method was performed on *Saccharomyces cerevisiae* histone H3 and FLAG-H2B, carried through the DNA linear amplification method

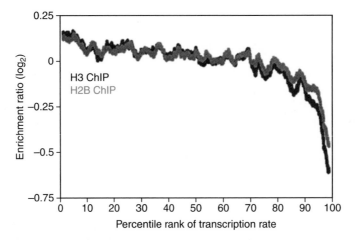

Figure 4. Example of ChIP–chip results.
Results from ChIPs of *S. cerevisiae* histone H3 and FLAG-H2B, plotted by intergenic genomic loci as a function of the percentile rank of the transcription rate in the downstream genes (14). An inverse relationship can be seen between nucleosome occupancy (as represented by enrichment for histone H3 and H2B) and promoter strength, as indicated by the transcription rate. These ChIP samples were amplified with the DNA linear amplification protocol presented in this chapter. Reprinted with permission from (14).

presented in this chapter, reverse transcribed, labeled, and hybridized on to *S. cerevisiae* genome arrays consisting of intergenic regions. This was plotted against the transcription rate of downstream genes; one can observe the inverse relationship between nucleosome occupancy and promoter strength.

4. TROUBLESHOOTING

These protocols routinely work well when the samples used are within the recommended starting amount range, and particularly when users new to these protocols use the controls described in section 2.5. Poor aRNA yield is the most frequently encountered problem. Occasionally, one may encounter more subtle problems, either stemming from the nature and composition of the template DNA used, or in ensuring compatibility with downstream applications such as microarrays.

4.1. Poor amplification yield

If aRNA yield is poor, examination of the controls can quickly pinpoint the likely cause of the problem.

4.1.1. RNase contamination

The IVT control provides a good way of determining whether there are any problems associated with handling RNA and maintaining an RNase-free environment. If the IVT control produces a poor yield, this may be due to contamination with RNases. This can be confirmed by running a 2% non-denaturing agarose gel in Tris-acetate-EDTA (TAE) and ethidium bromide. An RNase-contaminated IVT sample will yield a smear of low-molecular-weight material. If RNase contamination is determined to be the cause, ensure that aerosol-barrier, RNase-free pipette tips are used, and that working surfaces are treated with RNaseZap (Ambion) or other RNase-decontaminating agents. This is particularly important if working with ChIP samples. A way to verify that the RNase A used to digest RNA carried over from the ChIP is completely eliminated is to use the pTRI-Xef linear plasmid supplied with the Ambion IVT kit, add RNase A to the amount typically used for post-ChIP RNA digestion, and then carry it through three successive cleanups using *Protocol 6.*

4.1.2. Suboptimal IVT conditions

If there is no RNase contamination detected, either via the A_{260}/A_{280} ratio from a UV absorbance reading of the sample, or from analyzing the sample by gel electrophoresis, it is likely that there are problems with the IVT reaction conditions:

- The NTP mix may have gone through too many freeze–thaw cycles. As described in *Protocol 4*, NTPs are very sensitive to freeze–thaw cycles and each one decreases the yield. If this is the case, use a fresh IVT kit and aliquot the NTP mix (as described in *Protocol 4*, note [a]; into more than four aliquots if necessary) before use.
- Excessive evaporation of the reaction volume may have occurred during incubation. *Protocol 4* describes the proper incubation conditions for the IVT reactions. These conditions have been designed to limit evaporation and vapor volume during the long incubation period. Follow the conditions described in *Protocol 4*. Using mineral oil is not recommended, since it may interfere with either the IVT reaction (see *Protocol 4*) or the aRNA cleanup (see *Protocol 5*), or both.

You may also consult the troubleshooting section in the manufacturer's manual accompanying the IVT kit.

4.1.3. Poor yield with the positive-amplification control (*Protocols 2 and 3*)

This is likely to occur if the considerations mentioned in section 2.5 were not followed.

- If the template DNA has a large proportion of 3′ recessed ends, this template will not tail efficiently. A fill-in reaction using Klenow enzyme is recommended. ChIP samples from sheared genomic DNA typically have approximately half of their ends 3′ recessed; this correspondingly reduces the yield by half. Double the starting amount or use Klenow fill-in to obtain yields comparable to those indicated in *Table 2*.

- If the template DNA has a large proportion of 3′ phosphate groups, the template will not tail efficiently. Treat the template DNA with an alkaline phosphatase, as suggested in *Protocol 1*. Note that some DNA fragmentation methods such as restriction digestion may leave behind 3′ phosphate groups. Because this information is not always readily available for a given restriction enzyme, try *Protocol 1* on the template DNA and see whether this solves the yield problem.

- Check the recovery yield of the template DNA at the end of *Protocol 6* (Qiagen MinElute column cleanup). From 50 ng of input DNA, a yield of 50–80% is typical. If the yields obtained are significantly less than this, verify that the column membrane is completely wetted by Buffer EB during step 6 of *Protocol 6*. Furthermore, verify that the columns are viable – starting from 2004, Qiagen required that the MinElute columns be stored at 4°C when not in use.

4.1.4. Formation of template-independent product

The amplification product may contain a substantial amount of template-independent product when the mass ratio of T7 primer to template DNA significantly exceeds 5:1. This can easily occur when the starting amount of the template DNA is significantly overestimated. An example of this template-independent product is shown in *Fig. 3* (*lane 4*). We speculate that excess T7 primer during *Protocol 3* produces primer dimers, which yields an IVT template that produces the band of low-molecular-weight material shown in *Fig. 3*. This IVT side reaction diminishes the yield of the true amplification product. To prevent this from happening, follow *Table 1* in *Protocol 3*, which indicates the proper volumes and concentrations of T7 primer and template DNA to use for this step. If necessary, follow the recommendations discussed in section 2.4 for accurate measurement of the starting amount of template DNA.

4.2. Poly(A) tracts in the amplified RNA

The TdT step (see *Protocol 2*) produces poly(dT) tails within a size range of 20–40 nt. While we speculate that overhanging 3′ regions of the poly(dT) tails in the second-strand synthesis step (see *Protocol 3*) are removed via the 3′ exonuclease activity of the Klenow fragment enzyme, there exists the possibility that the T7 primer may anneal in such a way that the 3′ anchor of the primer (denoted by base designation '[B]' in *Fig. 2*) may not be base paired, and that the rest of the poly(dA) region of the primer is base paired anywhere along the length of the poly(dT) tail. The resulting size distribution may thus be larger than the original template, usually in the order of 20–40 bp, and would appear as a gel shift and a broadening of the gel bands corresponding to that size range. This issue has not been examined carefully because the potential variability in tail length does not appear to affect amplification efficiency and fidelity when used on microarrays. However, this issue may be important for consideration of applications that are sensitive to this potential variability in tail lengths and to the poly(A) tracts that will appear in the final amplification product.

4.3. DsRNA formation in the amplified RNA product

Theoretically, dsRNA can conceivably form from the amplification products, since aRNA based on both strands of the original template is produced. However, we have not tested whether dsRNA actually forms under the conditions outlined in this method. For spotted microarray experiments, if DNA probe produced from the aRNA product undergoes some degree of self-hybridization, the end result could potentially be a decrease in net signal intensity or compression of the dynamic range in the ratiometric data obtained, or both. We have observed this in one case with a yeast open reading frame microarray where, in a low-complexity mixture containing less than 300 unique DNA species, amplification of both strands compressed the dynamic range by ~60–70% when compared with that obtained with a single-strand amplification (Rebecca Butcher, Harvard University, MA, USA, personal communication). We were able to make this determination via single-strand amplification because the starting material already had conserved sequences that were different on each end of the amplicon. Note that the protocols as described in this chapter normally do not provide that opportunity.

We believe this issue, however, should not significantly impact on most studies that amplify highly complex mixtures of DNA, such as randomly fragmented genomic DNA. The amplification fidelity and signal quality have already been demonstrated to be at least as good as direct, unamplified Klenow labeling and better than R–PCR (6). We speculate that in a microarray hybridization, a DNA probe synthesized from a high-complexity mixture of aRNA (such as from amplification of sheared genomic DNA) is less likely to be affected by probe self-hybridization than probe synthesized from a low-complexity mixture (such as from amplification of a transcription factor ChIP that localizes to a small number of locations within the genome). The reason is that, during the hybridization process, a given probe strand in a highly complex mixture is more likely to hybridize to its complementary target on the microarray than to its complementary probe strand floating free in solution. This is because the complementary target on the microarray is fixed in location, while the complementary probe strand is free-floating and migrating throughout the hybridization solution. Thus, we believe that only in the case where probe composition is of low complexity should the user be concerned about probe self-hybridization.

If dsRNA does form in significant proportions, it may also reduce the efficiency and yield of reverse transcription either by slowing down the reverse transcriptase enzyme in the reverse transcription step or by causing insufficient denaturing of the dsRNA, leading to less efficient primer annealing. To compensate, we suggest using 50% more RNA than the amount typically used for microarray probe labeling. Nevertheless, we have not found it necessary to investigate carefully the potential impact of aRNA self-hybridization on reverse transcriptase efficiency, since high amplification fidelity is typically obtained. Furthermore, we have found in many cases that foregoing this increase will still yield a lower but usable net signal intensity for most spotted microarray hybridizations.

Acknowledgements

C.L.L. is supported by a Graduate Research Fellowship from the National Science Foundation. S.L.S. is an investigator at the Howard Hughes Medical Institute. B.E.B. is supported by a K08 Development Award from the National Cancer Institute. This work was supported by a grant from the National Institute for General Medical Sciences.

5. REFERENCES

★★ 1. Buck MJ & Lieb JD (2004) *Genomics*, **83**, 349–360. *– An excellent review of the ChIP–chip method.*

2. Ren B, Robert F, Wyrick JJ, *et al.* (2000) *Science*, **290**, 2306–2309.

3. Bohlander SK, Espinosa R, III, Le Beau MM, Rowley JD & Diaz MO (1992) *Genomics*, **13**, 1322–1324.

4. Gerton JL, DeRisi J, Shroff R, Lichten M, Brown PO & Petes TD (2000) *Proc. Natl. Acad. Sci. U. S. A.* **97**, 11383–11390.

5. Iyer VR, Horak CE, Scafe CS, Botstein D, Snyder M & Brown PO (2001) *Nature*, **409**, 533–538.

★★★ 6. Liu CL, Schreiber SL & Bernstein BE (2003) *BMC Genomics*, **4**, 19. *– Original publication describing the DNA linear amplification method presented in this chapter.*

★ 7. Baugh LR, Hill AA, Brown EL & Hunter CP (2001) *Nucleic Acids Res.* **29**, e29. *– First detailed description of elimination of template-independent amplification product.*

★ 8. Phillips J & Eberwine JH (1996) *Methods*, **10**, 283–288. *– Description of the first RNA linear amplification method.*

9. Wang E, Miller LD, Ohnmacht GA, Liu ET & Marincola FM (2000) *Nat. Biotechnol.* **18**, 457–459.

★ 10. Marko NF, Frank B, Quackenbush J & Lee NH (2005) *BMC Genomics*, **6**, 27. *– The background introduction provides an excellent summary of the state of the RNA amplification field at the time of publication of this chapter.*

11. Sambrook J & Russell DW (2001) *Molecular Cloning, a Laboratory Manual.* Cold Spring Harbor Laboratory Press, Cold Spring Harbor, New York.

12. Martin CT, Muller DK & Coleman JE (1988) *Biochemistry*, **27**, 3966–3974.

13. Lage JM, Leamon JH, Pejovic T, *et al.* (2003) *Genome Res.* **13**, 294–307.

14. Bernstein BE, Liu CL, Humphrey EL, Perlstein EO & Schreiber SL (2004) *Genome Biol.* **5**, R62.

15. Bernstein BE, Kamal M, Lindblad-Toh K, *et al.* (2005) *Cell*, **120**, 169–181.

16. Humphrey EL, Shamji AF, Bernstein BE & Schreiber SL (2004) *Chem. Biol.* **11**, 295–299.

17. Lee CK, Shibata Y, Rao B, Strahl BD & Lieb JD (2004) *Nat. Genet.* **36**, 900–905.

CHAPTER 8

Multiple displacement amplification of genomic DNA

Roger Lasken

Center for Genomic Sciences, Allegheny-Singer Research Institute, West Penn Allegheny Health System, 320 North East Ave. Pittsburgh, Pennsylvania 15212-4772, USA

1. INTRODUCTION

High-throughput genotyping assays have revolutionized genetics but require large quantities of genomic DNA. DNA has usually been obtained from blood and other specimens using extraction and purification methods. However, use of amplified DNA for genotyping has increasingly become an attractive alternative. WGA methods are now reliable enough to produce high-fidelity sequences. The multiple displacement amplification (MDA) method (1–6) generates hundreds of micrograms of DNA from minute specimens such as a few microlitres of blood (6) and even fingerprints (7). MDA is capable of replicating an estimated 99.8 % (8) of the genome with relatively low amplification bias in the representation of different regions of the genome (2, 6). It is the first WGA method to generate high-molecular-weight DNA (1, 2) of 2 kb to >100 kb in length compared with PCR-based methods where products are only a few hundred base pairs and there is greater bias in sequence representation (2).

MDA is frequently used for the restoration of depleted collections of purified DNA. Even trace amounts of remaining DNA can serve as template for WGA. MDA is also a valuable method for DNA sample preparation from clinical specimens (6, 9). The amplified DNA is pure enough to add directly to most genotyping assays, bypassing bead or cartridge cleanup steps. Furthermore, the accuracy of PCR detection can be improved by use of amplified DNA compared with PCR directly from crude specimens because the amplified DNA is enriched relative to contaminants that inhibit *Taq* DNA polymerase, such as hemoglobin and nucleases, which degrade primers, template, and PCR products. The abundant amplified DNA can also allow the use of as much DNA as desired in genetic assays. Insufficient DNA template is a common cause of higher error rates in all single nucleotide polymorphism (SNP) assays including hybridization, sequencing,

Whole Genome Amplification: *Methods Express* (S. Hughes and R. Lasken, eds.)
© Scion Publishing Limited, 2005

endonuclease, and ligation-based assays. Having a large supply of DNA may also allow more replicate assays as a means of improving accuracy, as well as allowing testing for many genetic loci from a single specimen.

MDA gives a particularly complete WGA that behaves like the original genomic DNA for many applications. The high-molecular-weight DNA makes MDA the preferred method for use in restriction fragment length polymorphism (RFLP) detection (2, 10), analysis of chromosomal translocations and gene rearrangements (10), and DNA sequencing (1, 3, 8, 11, 12). The φ29 DNA polymerase used in MDA also has a very high fidelity (4, 13) resulting in extremely accurate genotyping of SNPs and point mutations (2, 6, 8, 11, 14, 15). MDA was reported to provide a higher power of genotype discrimination for SNPs by mini-sequencing than improved primer-extension pre-amplification (I–PEP) (11). Complete genomic DNA amplification with low amplification bias depends on starting with sufficient copies of the DNA template (see section 2.2). The special use of MDA for genomic DNA amplification from single cells is considered in other chapters. While greater amplification bias occurs when starting from only a single cell, many potentially valuable applications are being explored (see Chapter 9 for MDA protocols for single bacterial cells and applications to unculturable species, and Chapter 11 for single human cells and applications to pre-implantation genetic diagnostics).

1.1. The challenge of amplifying an entire genome

It is no simple matter to replicate the 3.2 billion nucleotides of the human genome. Of course, the oldest 'WGA' processes evolved in living cells. As yet, no *in vitro* reaction can match the accuracy and completeness with which cells conserve their genetic information during DNA replication. The complexity of cellular processes illustrates the challenges. More than 20 different purified proteins are required to reconstitute replication of dsDNA *in vitro* (16). Origin-binding proteins initiate separation of the two strands forming a replication bubble (see *Fig. 1a*). DNA helicases expand the replication bubble and proceed along the DNA template ahead of the DNA polymerases. Helicases use the energy of ATP hydrolysis to separate the two strands at the replication forks. ssDNA-binding proteins are involved in stabilizing the separation of the two strands and primases synthesize RNA primers. Additional proteins tightly clamp the polymerase on to the primed DNA template for processive DNA synthesis of strands many kilobases in length. Unfortunately, for biotechnology applications, these reconstituted reactions have been too complex to make use of. While complete chromosomal DNA replication may some day become a routine laboratory technique, this goal has yet to be achieved.

PCR (17, 18) solves the problem of replicating dsDNA by melting the template at 95°C, annealing and extending primers at lower temperature, and cycling back to 95°C to melt the new double helix. The first WGA methods used PCR with either random primers in a method called primer-extension pre-amplification (PEP) (19) or degenerate primers in a method called degenerate-oligonucleotide-primed PCR (DOP–PCR) (20, 21).

(a)   **DNA replication in the cell**

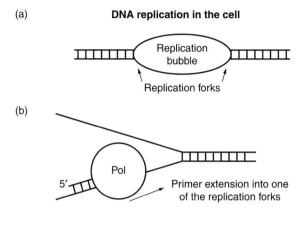

(b)

Primer extension into one
of the replication forks

(c)   **MDA reaction mechanism**

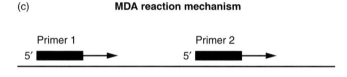

(d)

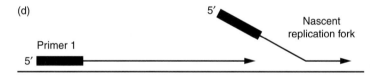

(e)

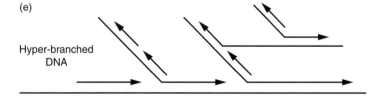

Hyper-branched
DNA

Figure 1. Replication of dsDNA.
(*a*) In the cell, DNA replication proteins are required to pull apart the two stands of the
double helix, creating a replication bubble with replication forks at either end. (*b*) The
DNA polymerase moves along the template on the leading strand, invading the replication
fork as the complementary DNA strand is displaced. In the cell, strand-displacement
synthesis requires the actions of DNA helicases and ssDNA-binding proteins to separate
the two DNA strands. In MDA, φ29 DNA polymerase is able to invade a replication fork
without the aid of other proteins. (*c*) As φ29 DNA polymerase extends a primer (primer
1), other downstream primers and their elongation products are in its path (primer 2). (*d*)
φ29 DNA polymerase is capable of establishing new replication forks (nascent replication
forks) when it encounters the 5′ end of these downstream products. (*e*) As ssDNA is
displaced, it in turn becomes available for yet more primers to anneal to. Amplification
becomes exponential by a hyper-branching mechanism.

1.2. Reaction mechanisms and enzymology of MDA

MDA is the first WGA method that is not based on PCR. It is an isothermal reaction carried out at 30°C. It achieves replication of dsDNA in a similar way to the cellular process in which the DNA polymerase invades a replication fork (see *Fig. 1b*). MDA utilizes a very unusual DNA polymerase from the bacteriophage φ29 to extend random hexamer primers. The key to MDA is that the φ29 DNA polymerase is able to invade the replication fork efficiently without the aid of a helicase. Its extremely tight binding to template and so-called strand-displacement activity allow it to 'push' its way into a replication fork, displacing the complementary strand of the template as it proceeds (see *Fig. 1b*).

In the initial step of MDA, genomic DNA is denatured either by treatment with KOH or by heating at 95°C. Random hexamer primers and the φ29 DNA polymerase are then added to the ssDNA and primer extension begins (see *Fig. 1c*). The amplification becomes exponential because of the ability of φ29 DNA polymerase to establish new replication forks when it encounters regions of dsDNA. As the DNA polymerase is extending a primer (see *Fig. 1c*, primer 1) on a ssDNA template, other downstream primers and their extension products are in the path of the polymerase (see *Fig. 1c*, primer 2). Some DNA polymerases, such as *Escherichia coli* DNA polymerase III, have been shown to pass over a double-stranded region and proceed downstream without initiating strand-displacement synthesis (22). However, φ29 DNA polymerase is proficient at invading the duplex at the 5′ end of the downstream primer and establishing a new replication fork (see *Fig. 2d*, nascent replication fork). Furthermore, the displaced strand becomes available for annealing of yet more random primers. φ29 DNA polymerase is thought to establish a 'hyper-branched' structure (23), extending additional primers as it concurrently displaces downstream products (see *Fig. 1e*). Hyper-branching creates a highly exponential amplification. For example, more than a billionfold amplification is achieved from the genome of a single bacterial cell in an overnight MDA reaction (24). About 5 fg of genomic DNA in a single bacterium yields about 25 µg of amplified DNA.

For MDA, the critical characteristics of φ29 DNA polymerase are its exceptionally tight binding to DNA and the related properties of processivity and strand-displacement synthesis. φ29 DNA polymerase has a processivity of about 70 000 nt (25) meaning that, on average, it adds 70 000 nt each time it binds to the primer/template. For comparison, the large fragment of *E. coli* DNA polymerase I (Klenow fragment) has a processivity of only about 10 nt. A typical processivity assay demonstrates the appearance of these short DNA products (see *Fig. 2a*, arrow) under conditions of low polymerase concentration and large excess of primed DNA template. As more Klenow fragment is added, longer products appear, but they have resulted from the repeated binding and dissociation of polymerase. By contrast, limiting amounts of φ29 DNA polymerase generate enormous DNA products that are beyond the 23 kb resolution of the denaturing agarose gel shown (see *Fig. 2b*, arrow). While high processivity contributes to the robust strand-displacement synthesis, the dynamics of the hyper-branching mechanism (see *Fig. 1e*) also presumably determine the average length of the

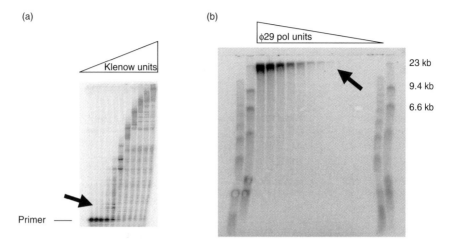

Figure 2. Processivity assay.
A primer labeled at its 5′ end with ^{32}P was annealed to M13 ssDNA and extended with varying amounts of DNA polymerase. (*a*) A fourfold serial dilution of Klenow fragment (ranging from 0.00008 to 20 units, and a control lacking DNA polymerase at the far left) was added and allowed to extend the primers. Reaction products were resolved on a polyacrylamide gel and visualized by autoradiography. Note that at limiting polymerase concentration, most of the primers remain unextended. This ensures that the few primer extensions that have occurred are the result of a single binding of the polymerase to the primer/template. The true processivity of only about 10 nt is observed (arrow). Longer reaction products at higher concentrations of polymerase result from the polymerase rebinding many times to the DNA. (*b*) Processivity of φ29 DNA polymerase with reaction products resolved on a denaturing alkaline agarose gel. A twofold serial dilution of φ29 DNA polymerase ranging from 0.01 to 10 units (from right to left, with a control lacking DNA polymerase at the far right) was used. The DNA products (arrow) exceed the 23 kb resolution of this gel, indicated by the λ *Hind*III ladder (Invitrogen) on the right.

amplified DNA products in MDA. The average length is about 12 kb (2) when run on a denaturing gel so that the DNA is resolved as the single-stranded products. As the amplification proceeds, annealing of new primers internally on previously displaced product strands should act to limit the average product length generated.

The ability of φ29 DNA polymerase to bind tightly to the DNA template is demonstrated in a primer-extension experiment (see *Fig. 3*). The polymerase was allowed to bind to a synthetic oligonucleotide primer/template in a pre-reaction mix. An M13 DNA trap was then added to the reaction to capture any DNA polymerase that dissociated from the primer/template and prevent it from rebinding. The mix was incubated for varying times to determine the rate at which the polymerase dissociated from the primer/template. Finally, dNTPs were added to the mix. The number of primers that were extended revealed how many of the primer/templates still had associated DNA polymerase. The half-life of the φ29 DNA polymerase/DNA complex ranged from about 1 to 13 min depending on the

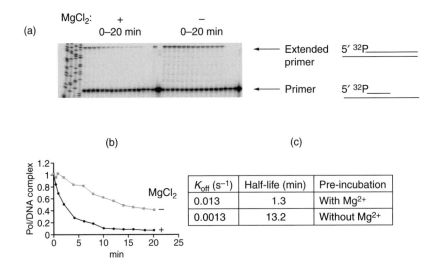

Figure 3. Binding of φ29 DNA polymerase to DNA.
φ29 DNA polymerase was allowed to bind to a pre-annealed complex of 5′ ^{32}P-labeled synthetic primer (20 nt) annealed to template (70 nt). An M13 DNA trap was then added to the mix to capture any polymerase that dissociated from the DNA. The polymerase was allowed to dissociate from the DNA for the number of minutes indicated in (*a*) and (*b*). Finally, dNTPs were added to extend the primers to the end of the template. Primer extension was assessed by resolution on a polyacrylamide gel and autoradiography. (*a*) The amount of primer that is extended by 41 nt to the end of the template reveals how many of the primer/templates still had associated DNA polymerase. The primer anneals such that there is a 9 nt 3′ overhang on the template and the DNA polymerase will run off the template after adding on 41 nt. (*b*) Primer extension was quantified using a Molecular Dynamics Storm PhosphorImager and ImageQuant software supplied by the manufacturer. (*c*) The half-life of the φ29 DNA polymerase/DNA complex ranged from about 1 to 13 min depending on the incubation conditions.

incubation conditions. These results are in reasonable agreement with the processivity data since the polymerase can synthesize about 50–200 nt/s or about 3000–150 000 nt in one DNA binding event of 1–13 min. More investigation will be required to understand fully the relationship between DNA binding, processivity, and MDA reaction characteristics. Interestingly, φ29 DNA polymerase bound more tightly to the DNA in the absence of $MgCl_2$, suggesting a strong ionic effect on the dissociation constant (K_D).

Establishment of a nascent replication fork and subsequent strand-displacement synthesis can be observed in a simple experimental system in which a ^{32}P-labeled primer is extended on an 80 nt circular template (see *Fig. 4*). Extension of the primer once around the circle results in a DNA product of 80 nt (see *Fig. 4*, nicked circle). All subsequent elongation requires the polymerase to invade the 5′ terminus of its own primer and enter into a rolling-circle mode of DNA synthesis. The replication fork established is essentially identical to those in the hyper-branching model (see *Fig. 1d* and *e*). The rate of strand-displacement

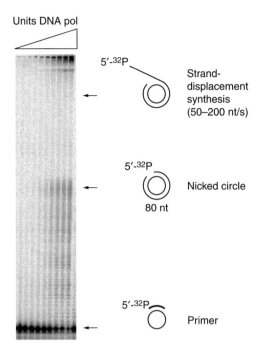

Figure 4. Strand-displacement synthesis.
φ29 DNA polymerase was used to extend a 5′ ³²P-labeled primer annealed to an 80 nt circular DNA template. Reaction products were resolved on a polyacrylamide gel. The position of the unextended 20 nt primer is indicated at the bottom of the gel (primer). Following the addition of φ29 DNA polymerase (0, 0.01, 0.02, 0.08, 0.16, 0.6, 1.2, 2.5, 5, and 10 units in lanes 1–10, respectively), the primer was extended once around the circle to a length of 80 nt (nicked circle). Subsequent elongation requires the DNA polymerase to invade a downstream DNA 5′ end, establishing a replication fork and initiating strand-displacement synthesis.

synthesis measured over various time courses (data not shown) ranged from about 50 to 200 nt/s.

Efficient MDA and relatively even coverage of the genomic DNA template depends on maintaining high concentrations of DNA polymerase and primer throughout the reaction. However, φ29 DNA polymerase rapidly degrades normal primers with its 3′→5′ exonuclease proofreading activity (1). The proofreading activity contributes to high-fidelity replication by excising nucleotides misincorporated by the DNA polymerase activity. Unfortunately, it also avidly degrades single-stranded oligonucleotides such as the random hexamer primers (see *Fig. 5*, exo sensitive). The problem was solved by using random hexamers that are protected from degradation (see *Fig. 5*, exo resistant) with two phosphorothioate linkages at the 3′ terminus (5′-NpNpNpNp^{S}Np^{S}N-3′). With the primers protected from degradation, the amplification could occur (see *Fig. 6*) (1).

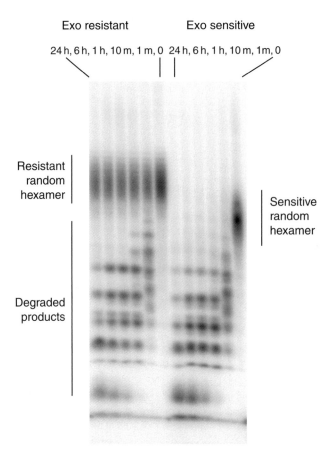

Figure 5. Protection of random primers from degradation by the ϕ29 DNA polymerase-associated 3′→5′ exonuclease activity.
5′ ^{32}P-labeled random hexamer primers were subjected to 0.5 units/μl of ϕ29 DNA polymerase for the times indicated and resolved on a 25% polyacrylamide gel. The exonuclease (exo)-resistant primers remained full length for 24 h, while the sensitive primers were degraded within 1 min. The exo-resistant random hexamer primers (synthesized by Integrated DNA Technologies) had a sulfur atom substituted for oxygen on the phosphate group of the two 3′-terminal nucleotides (5′-NpNpNpNp^{S}Np^{S}N-3′). The DNA polymerase activity can extend the primers; however, its 3′→5′ exonuclease activity is blocked. The full-length primers (time 0) appeared as a fairly broad family of bands due to the different mobilities of the random sequences. Note that the exo-resistant primers had a slower mobility than the exo-sensitive primers due to the sulfur atoms, which are almost neutral (30).

2. METHODS AND APPROACHES

2.1. Source of DNA template

MDA is frequently used for increasing the supply of purified genomic DNA already present in stored collections. The quality of the stored DNA to be used as template

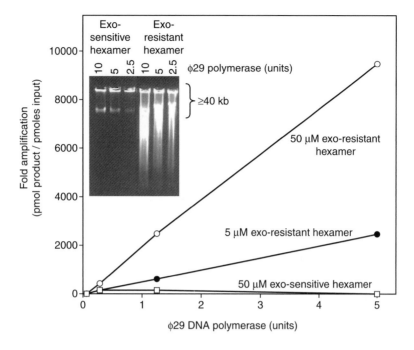

Figure 6. MDA with exonuclease-sensitive or exonuclease-resistant random primers.
MDA reactions in a volume of 20 µl contained 1 ng of M13 DNA as template with
exonuclease (exo)-resistant or exo-sensitive primers and φ29 DNA polymerase at the
concentrations indicated. DNA synthesis was quantified by incorporation of radioactive
nucleotide. Reprinted from (1) with permission of *Genome Research*, Cold Spring Harbor
Laboratory Press.

for MDA is critical in obtaining high-quality amplified DNA. MDA can also be used
for amplification of plasmids (1) and bacterial artificial chromosome libraries (26).
MDA is also used for DNA amplification directly from clinical and other biological
specimens such as blood, buffy coats, buccal swabs, dried blood (6, 27),
fingerprints (7), plasma, serum, cultured cells and tissues (27), and urine (Paul
Wood, Genomics and Proteomics Core Laboratory, University of Pittsburgh, USA,
personal communication). The highest-quality amplified DNA is often obtained by
direct addition of biological specimens to the MDA reaction. Extraction and
purification of DNA from clinical specimens, such as blood and buccal swabs, prior
to use in MDA not only requires more steps, but also introduces more
opportunities for DNA loss, damage, and contamination. MDA yields highly
enriched DNA that usually does not require bead- or cartridge-based cleanup or
purification steps prior to use in downstream assays. The amplification enriches
the DNA relative to cellular debris, inhibitors including nucleases, hemoglobin,
anti-coagulants added to blood samples (e.g. EDTA, citrate, acid–citrate–dextrose,
and heparin), growth media, or other contaminants found in crude clinical or field
samples.

2.2. Amount of biological specimen required as DNA template source

Use 0.5 μl of blood or buffy coats, or 0.5 μl of cell material (>600 cells/μl) such as cultured cells. In general, the gently lysed cells provide high-molecular-weight DNA template for amplification, whereas extracted and purified DNA template has undergone more shearing and damage. Therefore, 10–100 ng of purified DNA is recommended (see sections 3.1 and 3.3.2), which represents about 3000–30 000 human genome copies. For tissue sections, buccal swabs, and dried blood spots, a varying quality and amount of genomic DNA is obtained. One suggested approach, when possible, is to use larger-volume MDA reactions of 500 μl for these specimens, allowing more of the cell lysate to be added to the reaction (27). The recommended MDA volume for plasma and serum lysates where only trace amounts of DNA may be present is 750 μl (27).

2.3. Source of MDA reagents

MDA protocols currently in the literature may vary (1–3, 5). MDA reagents can be obtained from several vendors (Qiagen; Amersham Biosciences; New England Biolabs; Epicentre). The φ29 DNA polymerase purity and concentration are critical to achieve peak performance in MDA. For convenience, commercial MDA-based WGA kits can be purchased (Qiagen; Amersham Biosciences) and are validated by their manufacturers for optimal reaction conditions. Peak performance for replication fidelity, complete genome coverage, and low amplification bias may depend on careful quality control of MDA reagents. This is an advantage of pre-assembled WGA kits.

3. RECOMMENDED PROTOCOLS

A typical MDA protocol (27), based on alkaline denaturation of template, is given here. The general requirements for MDA are:

1. An appropriate source of DNA template (see section 2.1).
2. A means of denaturing the DNA template either with alkaline (2, 6) or with high-temperature (95°C) (1) treatment.
3. φ29 DNA polymerase, reaction buffers, and dNTP substrate to support DNA synthesis (see section 2.3).

3.1. MDA reaction

For WGA using purified human DNA, it is recommended that at least 10 ng of DNA template be added (about 3000 copies of the human genome). When required, less template can also be amplified; however, some loss of sequence from the amplified DNA is possible. If the DNA is suspected of being partially degraded, the addition of more DNA (30–100 ng) may improve results.

Protocol 1

MDA

Equipment and reagents

■ Denaturation solution (400 mM KOH; 10 mM EDTA). Include 100 mM dithiothreitol (DTT) for cell lysis for blood, cultured cells, and other biological specimens. It is recommended that DTT be omitted for buccal swabs as better results are sometimes observed. The denaturation solution can be stored for 1 week at room temperature in a tightly closed bottle to prevent neutralization.

■ Neutralization solution (800 mM Tris-HCl (pH 4 unadjusted, for example, from Sigma))

■ ϕ29 DNA polymerase (Qiagen; Amersham Biosciences; New England Biolabs; Epicentre)

■ Nuclease-free water

■ 4× MDA reaction mix (27) (150 mM Tris-HCl (pH 7.5); 200 mM KCl; 40 mM MgCl$_2$; 80 mM ammonium sulfate; 4 mM each of dATP, dGTP, dCTP, and dTTP; 0.2 mM random hexamers protected by phosphorothioate modification on the two 3'-terminal nucleotides (5'-NpNpNpNp^{S}Np^{S}N-3'; synthesized, for example, by Integrated DNA Technologies) (1)

■ Microcentrifuge tubes or microtiter plates

■ Thermal cycler, water bath, or heating block

Methods

1. Carry out cell lysis (in the case of biological specimens) and alkaline denaturation by adding at room temperature 3 μl DNA (10–100 ng) or 3 μl biological specimen (see section 2.2) to 3.5 μl denaturation solution. Mix and incubate for 3 min at room temperature.

2. For neutralization of the DNA template (for both biological specimens and purified DNA templates), add 3.5 μl neutralization solution and mix. Use in the MDA reaction within 1 h.

3. For the MDA WGA reaction[a,b,c]:
   ■ Add 27 μl of H$_2$O to the 10 μl of denatured and neutralized DNA resulting from step 2.
   ■ Add 12.5 μl of 4× MDA reaction mix.
   ■ Add 0.5 μl of ϕ29 DNA polymerase and mix well. The final reaction volume is 50 μl.

4. Incubate for 6–16 h at 30°C[d].

5. Terminate the reaction by heating at 65°C for 3 min[e].

Notes

[a]This protocol was developed from the REPLI-g kit (Qiagen). ϕ29 DNA polymerase unit concentrations from other vendors may vary. Always use the manufacturer's recommended amounts.

[b]For 50 and 100 μl reaction volumes, assemble in 0.2 ml tubes or microtiter plate wells.

[c]When preparing multiple reactions, ϕ29 DNA polymerase can be added, on ice, to the 4× MDA reaction mix in multiples of 0.5 μl polymerase/12.5 μl 4× MDA mix. Add the polymerase to the 4× MDA reaction mix immediately before use. Add 13 μl of the polymerase/MDA mix to each solution of water (27 μl) and DNA (10 μl) and mix thoroughly (final reaction volume 50 μl). The polymerase is sufficiently stable while the MDA mix is at 4× concentration. Adding polymerase to a more dilute reaction mix is not recommended. It is critical to mix the polymerase/4× MDA reaction mix thoroughly as polymerase stocks are usually 50% glycerol and will settle in the bottom of the tube.

[d]Reaction termination between 8 and 16 h assures maximum yield. Reactions are conveniently incubated overnight for 16 h.

[e]This step is essential in order to inactivate the DNA polymerase and its associated 3'→5' exonuclease activity, which can degrade the amplified DNA if not inactivated.

3.2. Expected yield of amplified DNA and storage

The DNA yield from the reaction can be determined with a DNA quantification method specific for dsDNA such as PicoGreen (Molecular Probes). This is required since unused reaction primers and dNTPs are present and methods such as UV absorbance will not accurately reflect DNA yield. MDA reactions tend to be self-limiting with yields of about 0.8 µg DNA/µl ± 15%. Therefore, the yield of amplified DNA is limited by the reaction volume, with a 50 µl reaction generating about 40 µg, and a 10 ml preparative reaction yielding about 8 mg of DNA. It is a property of MDA reactions that similar yields of nonspecific DNA are generated when no DNA template is added to the reaction. This nonspecific DNA is thought to be derived from reaction primers or trace amounts of contaminating DNA (2). For most applications, this will not be problematic. For example, PCR and genotyping assays simply gave no result when using MDA negative controls (11). However, in instances where MDA negative controls lacking input DNA template must not yield any DNA synthesis (as measured by PicoGreen, for example), this must be taken into consideration. Only subsequent genotyping, DNA sequencing, or other analysis will reveal whether the amplified DNA is specific. Determination of the number of micrograms of DNA yielded is still a useful quality control, since reduced yield is diagnostic for (i) the presence of inhibitors in specimens, such as excess hemoglobin, or nucleases, which degrade reaction products; (ii) the loss of φ29 DNA polymerase activity where the enzyme has exceeded its shelf life; or (iii) experimental error in adding the correct reaction components.

DNA yield tends to plateau in the reaction after about 6–8 h. Using a thermal cycler reduces evaporation, and allows programming of the 30°C incubation, the 65°C termination step, and storage at 4°C until the sample is retrieved from the thermal cycler. Longer incubation times at 65°C are required for terminating the reaction for large preparative reactions in order to bring them to a sufficient temperature to inactivate the polymerase. For example, a 10 ml preparative reaction carried out in a Falcon tube requires 10 min in a 65°C water bath. Amplified DNA can be stored at –20°C. For long-term storage, it should be aliquoted into multiple tubes to reduce degradation from freeze–thaw cycles and stored at –80°C.

3.3. Characteristics of amplified DNA and reaction optimization

3.3.1. Expected genome coverage of the amplified DNA and performance in genotyping assays

The robust activity of the φ29 DNA polymerase and the high random primer concentrations of 50 µM result in almost complete coverage of the genome. Of 2320 SNPs assayed throughout the human genome at Illumina Inc. (15), all were present in the amplified DNA (with no significant difference in call rate between amplified and unamplified DNA). The high-fidelity φ29 DNA polymerase (4, 13) resulted in accurate genotyping with >99.8% concordance with unamplified genomic DNA. Another study found 99.95% accuracy for 3814 genotyping calls and concluded that most of the errors were due to the genotyping assay and not the MDA (14). It has been reported that regions near the telomeres are underrepresented (5). However, it was estimated that 99.82% of the human genome was present in amplified DNA based on array probing of >10 000 SNP alleles (8). Direct sequencing of 500 000 bp of the amplified DNA yielded the same error rate of 9.5×10^{-6} as paired unamplified DNA. In one of the most thorough studies to date (11) using a DNA sequencing-based assay, MDA was preferred to I–PEP based on genotyping success, signal-to-noise ratio, power of discrimination between homozygous and heterozygous SNP genotypes, yield, and authenticity of allele representation.

Another advantage of the MDA method is the low amplification bias among different sequences. Of 47 single-copy loci tested, one on the p and q arm of each human chromosome, all were still represented at between about 0.5 and 3 copies per genome equivalent in the amplified DNA (6) (see *Fig. 7*). While some amplification bias had occurred over this sixfold range from 0.5 to 3 copies per genome, this was a large improvement over the PCR-based WGA methods, which have drastic variation in the representation of different sequences (2) (see *Fig. 8*).

3.3.2. Optimizing the MDA reaction for use in genotyping

It is important to minimize amplification bias created between the two alleles of an SNP or mutation. If one of the parental chromosomes is amplified

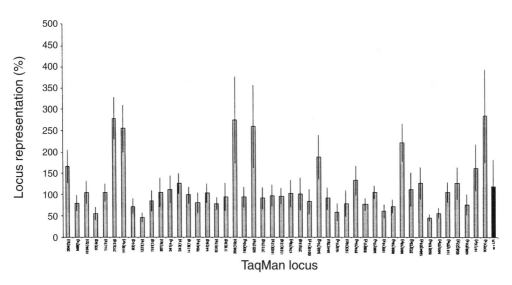

Figure 7. Amplification bias analysis of MDA by quantitative TaqMan PCR assays for 47 human loci. Loci representation is expressed as the percentage of locus copies in the amplified DNA relative to the starting DNA template. Each bar represents the average of amplifications of DNA from 44 different individuals. A value of 100% indicates that the sequence is present in the same copy number per genome in the amplified DNA as occurs in the starting genomic DNA template. The average of all 47 loci was 117% (far right bar). Reprinted from (6) with permission of *Genome Research*, Cold Spring Harbor Laboratory Press.

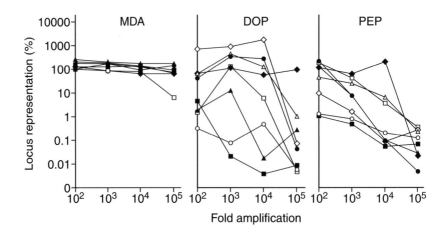

Figure 8. Effect of amplification on gene representation bias.
The relative representation of eight loci was determined using TaqMan quantitative PCR for DNA amplified by the MDA, DOP, and PEP methods as indicated. The x-axis represents the fold amplification compared with the starting DNA template; the y-axis is the locus representation. The results for eight loci are indicated as follows; ◇, CXCR5; △, connexin 40; □, MKP1; ○, CCR6; ◆, acidic ribosomal protein; ▲, CCR1; ■, cJUN; ●, CCR7. Reprinted from (2) with permission of the National Academy of Sciences USA.

preferentially over the other, this can result in miscalling of a heterozygous individual as homozygous. Maintaining a low error rate is important for linkage and association studies. In analysis of loss of heterozygosity (LOH) in cancer, artifact allele drop-out originating from the amplification will appear as natural LOH. Most critically, development of diagnostic methods for human patients will depend on very low error rates due to the potential severe consequences of both false-positive and false-negative genotyping. Published genotyping error rates (see section 3.3.1) demonstrate that MDA can be used successfully for many applications.

Using good-quality DNA template and adding a sufficient amount to MDA reactions is important for minimizing allele dropout. Use of ≥10 ng of genomic DNA template (about 3000 copies of the human genome) in MDA gives optimum genotyping performance (see *Fig. 9* in color section, standard gDNA). DNA of an individual heterozygous for the test locus was amplified in 84 replicate MDA reactions and then genotyped. In TaqMan SNP assays, heterozygous genotypes should appear as a tight population along the diagonal of the graph as is seen using DNA from MDA reactions that had started with 10 and 100 ng of DNA template (several points located near the x- and y-axes are homozygous controls). Use of only 1 ng or 100 pg of DNA template in the MDA results in scattering of the heterozygous population in the TaqMan assay and the potential for miscalling as homozygous. An older DNA sample, stored for a long period of time and known to be partially degraded, required more template in the MDA to achieve optimum genotyping (see *Fig. 9*, partially degraded gDNA). Clearly, a good strategy when using DNA of questionable quality is to add as much template to the MDA as possible. Strategies for salvaging partially degraded DNA collections are given below (see section 3.3.3). When biological specimens are available, such as blood and buccal swabs (6), it can be best to amplify directly from the specimens as this (i) assures high-molecular-weight template will be present; (ii) eliminates the work of DNA extraction prior to the MDA reaction (6); and (iii) makes the process easier to automate (3).

The scatter of PCR-based genotyping data observed (see *Fig. 9* in color section) strongly suggests that a minimum level of good-quality template is required to ensure equal amplification of both chromosomes. Real-time quantitative PCR was used directly to measure amplification bias created between the two parental alleles of an SNP. The two alleles were approximately equally represented in all 84 MDA reactions tested when starting from 100 ng of DNA template (see *Fig. 10*, MDA control template) with about 50% of allele 1 (*x*-axis) and 50% of allele 2. The standard deviation of only 1.5% was about the same as the 1.6% standard deviation for the unamplified template (see *Fig. 10*, control gDNA). When only 30 or 6 ng of DNA template was added to the MDA reaction, the standard deviation increased to only 2.9 and 3.7%, respectively (data not shown). Allele ratios were derived from a standard curve created by mixing together known amounts of homozygous DNA as described in *Fig. 10*. These experiments demonstrate how it is possible to replicate genomic DNA with such outstanding fidelity and accurate performance in the genome-wide SNP screens referenced above (see section 3.3.1). The high random primer and ϕ29 DNA polymerase concentrations produce robust amplification of any template present in the reaction. When an adequate number of genome copies is present as template, allele amplification bias is negligible. When older and partially degraded DNA template was used in MDA, allele bias was greater (see *Fig. 10*, MDA degraded template). A bell-curve distribution occurred in which both alleles were occasionally underrepresented at levels of only 10–20% relative to 80–90% of the other allele. These outliers could result in miscalling heterozygotes as homozygotes.

3.3.3. Restoration of partially degraded genomic DNA collections

The quality of stored DNA will depend on many factors including the quality of the initial DNA extraction, the age of the DNA, the number of freeze–thaw cycles it has undergone, and storage conditions. From experience, stored collections can often have a wider range of DNA concentrations than laboratories might expect. It is helpful to recheck the concentrations of a number of representative samples if sufficient DNA can be spared. Concentration is of particular concern for DNA extracted from specimens such as buccal swabs that are known to give variable yields (28). For partially degraded DNA, it is advisable to use as much template as possible in MDA reactions to ensure that sufficient high-molecular-weight template is present. For moderately degraded DNA, 100 ng is recommended (see *Fig. 9* in color section). If sufficient amounts are available, the genomic DNA can also be resolved by agarose gel electrophoresis to evaluate maximum and average length.

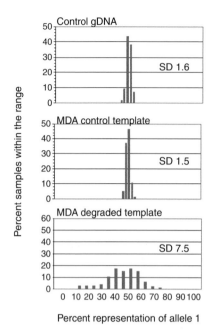

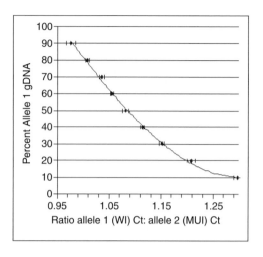

TaqMan standard curve at CYP2C 19*2 locus

Figure 10. Equal representation of both alleles in amplified DNA.

MDA reactions (84 replicates) were carried out with 100 ng (about 30 000 human genome copies) of a DNA sample (MDA control template) from an individual heterozygous for the CYP2C19 locus. Another 84 replicate MDA reactions were carried out from a heterozygous DNA template that was known to be partially degraded (MDA degraded template). Allele bias in MDA-amplified DNA was measured using the CYP2C19 Allelic Discrimination TaqMan Assay Kit (ABI). Real-time PCR was used to derive the relative representation of each allele (2, 6) based on the threshold cycle (Ct) at which detection occurred. The unamplified control genomic (g) DNA was also tested in the TaqMan assay (control gDNA). By definition, this unamplified heterozygous DNA is 50% allele 1 and 50% allele 2. MDA-amplified DNA samples were diluted 1:40 with sterile water and approximately 90 ng was added to the TaqMan assays. The ratio of allele 1 Ct (VIC) to allele 2 Ct (FAM) was determined from a standard curve generated by mixing together known amounts of homozygous DNA (standard curve). For example, when equal amounts of homozygous mutant and homozygous normal DNA were mixed together (simulating a heterozygote), a Ct ratio of about 1.08 was measured. The standard curve corrects for the unequal efficiencies between the PCRs for each allele. For every experiment, nine reactions containing gDNA heterozygous for the mutation were also assayed and the percentage of allele 1 was calculated and averaged, in order to normalize each experiment to the 50% point for equal amounts of the alleles. The standard curve was created by titrating gDNA in 10% increments (eight replicates) from 100% allele 1 (0% allele 2) to 0% allele 1 (100% allele 2). The TaqMan assay was performed as recommended by the manufacturer.

When sonicated DNA of different lengths was investigated, samples of at least 2–4 kb average length could be efficiently amplified (see *Fig. 11*). Shorter template was also amplified but with reduced efficiency. Reduced locus representation (i.e. lower copy number of the locus per microgram of amplified DNA) generally indicates that a higher percentage of nonspecific DNA synthesis is occurring. Nonspecific synthesis is thought to be derived from the random primers or trace amounts of contaminating DNA (2, 6). Many factors may influence MDA

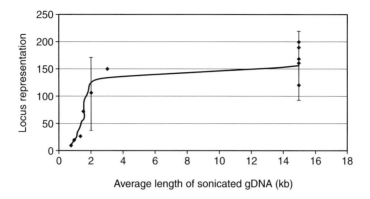

Figure 11. Effect of DNA template length on MDA performance.
Human DNA was sonicated to produce samples of different average length. The samples were then used as template for MDA reactions. Quantitative PCR using the TaqMan assay was carried out for a test locus to determine its representation in the amplified DNA. Use of DNA templates shorter than 2–4 kb in the MDA reactions resulted in a loss of locus representation in the amplified DNA.

performance. For the experimental test using sonicated DNA (see *Fig. 11*), the key determinant for optimal amplification may be that at least some small amount of high-molecular-weight DNA is still present, even when the average length is only 2–4 kb. The same may be true for gradually degrading DNA stored in archives. Samples obtained from one archive in long-term storage were also evaluated for average length on an agarose gel and then tested as template for MDA with subsequent quantitative TaqMan analysis for representation of a test locus (see *Fig. 12*). Again, partially degraded samples of only 2–4 kb in length (see *Fig. 12*, agarose gel, samples 1, 3, 4, and 6) resulted in lower representation of the test locus following amplification by MDA (see *Fig. 12*, TaqMan assay).

The highly processive φ29 DNA polymerase is expected to utilize longer DNA templates more efficiently. The hyper-branching mechanism itself (see *Fig. 1e*) might also require sufficient-length template to occur efficiently. Total DNA copy number available as template is also clearly important (see *Fig. 9* in color section), regardless of the length of the DNA, in order to represent both alleles of an SNP equally. Presumably, early in the MDA reaction, favored amplification of one allele over the other occurs on a random basis when there are very few template molecules present. As expected from this kind of stochastic process, neither locus is more likely to drop out. Rather, a bell-curve distribution occurs in which a few outliers of each allele are underrepresented (see *Fig. 10*, degraded template). Finally, there are many types of DNA damage that might occur in stored DNA or in field samples used for forensic or archeological studies that will also affect MDA performance.

Nevertheless, amplification of even moderately degraded DNA can be of value where no other alternative is possible. Typical genotyping assays, such as the PCR-based TaqMan assay, are highly specific for their target sequence. Therefore,

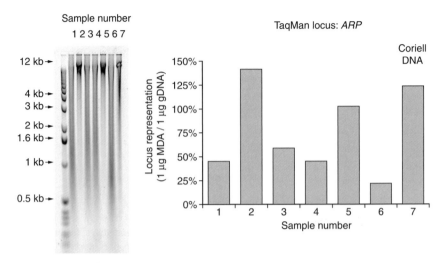

Figure 12. Poor locus representation when partially degraded human DNA is used as template for MDA.
Six samples that had been stored in a DNA collection (lanes 1–6) and control human DNA (Coriell; lane 7) were compared. Genomic DNA (500 ng) was loaded per well on a 0.8% agarose gel. The DNA samples were also used as template in MDA reactions and quantitative PCR was carried out for the *ARP* locus to determine its representation in the amplified DNA. Partially degraded DNA (lanes 1, 3, 4, and 6) identified on the agarose gel also generated amplified DNA with a poorer representation of the locus in the TaqMan assays.

amplified DNA from partially degraded template may still provide accurate genotyping. Caution should be used to evaluate carefully the quality of the amplified DNA for use in genotyping when the starting template was questionable. A quality-control assay has been developed to identify amplified DNA samples that are suitable for use in genotyping. Analysis of a few test SNPs is fairly diagnostic for the entire genome (29) and even samples extracted 10 years earlier could be recovered. Two human SNPs were selected as being particularly sensitive to any loss of DNA quality when resulting from use of degraded or low-concentration template in the MDA. Normal representation of these SNPs was generally correlated to comprehensive representation of the entire genome. Such quality-control assays can allow selection of an acceptable subset amplified from highly deteriorated collections and may well be justified when compared with the cost of obtaining new DNA samples from study subjects or abandoning the research entirely.

3.3.4. MDA from laser-capture microdissected tissue

The utilization of laser-capture microdissection in conjunction with WGA has enabled researchers to apply a wide range of genetic approaches to the study of small numbers of cells. The method is particularly useful in the analysis of cancer where small numbers of transformed cells can be separated from the normal cells that are also present in tumors. DNA extracted from 1000 laser-captured prostate cancer cells has been amplified by MDA and used for accurate comparative genomic hybridization (CGH) analysis (31). In addition, the use of pooled DNA from two or three separate amplifications has permitted the analysis of as few as 100 cells from prostate cancer laser-capture microdissected samples (32). The adoption of a statistical approach has allowed high-resolution profiling of copy number aberrations in colorectal adenomatous polyp cells by array CGH (33). This work took advantage of the small but reproducible differences in MDA efficiency (for example, see *Fig. 7*) to normalize data and allow detection of minor copy number changes.

4. TROUBLESHOOTING

- Reduced DNA yield when compared with the positive control (approximately 40 µg/50 µl expected). DNA synthesis failed in the MDA. There is a possible inhibitor in the DNA template or specimen. Clean up or dilute the DNA template source and re-amplify.
- Performance of the amplified DNA in genotyping or other applications is not optimum. Loci are missing or control heterozygous samples are miscalled as homozygous due to allele dropout. (i) The DNA template used may be degraded or of lower-than-expected concentration. Use intact template. Check DNA template concentration. Use a larger amount of template. (ii) The denaturation of template may have failed. Use fresh 5 M KOH stock solution to prepare the denaturation solution. (iii) Some sensitive genotyping or other applications may require DNA cleanup after MDA.
- Performance of the amplified DNA derived from blood samples is not optimum in genotyping or other applications. (i) A higher-than-normal concentration of heparin may have been used as blood anticoagulant. Dilute the heparin-treated blood up to five-fold using l× PBS. (ii) Ineffective lysis of cells. Use fresh 5 M KOH stock solution to prepare the denaturation solution.

5. REFERENCES

★★★ 1. **Dean FB, Nelson JR, Giesler TL & Lasken RS** (2001) *Genome Res.* 11, 1095–1099. – *First publication on MDA leading to Amersham Bioscience's TempliPhi product.*

★★★ 2. **Dean FB, Hosono S, Fang L, et al.** (2002) *Proc. Natl. Acad. Sci. U. S. A.* 99, 5261–5266. – *First publication on WGA using MDA. Demonstration of low amplification bias of MDA compared with other WGA methods.*

★★★ 3. **Detter JC, Jett JM, Lucas SM, et al.** (2002) *Genomics,* 80, 691–698. – *Excellent report on the use of MDA in high-throughput DNA sequencing and library construction from MDA products for bacterial genomic sequencing.*

★★★ 4. **Nelson JR, Cai YC, Giesler TL, et al.** (2002) *BioTechniques,* **Suppl.**, 44–47. – *Establishment of an error rate of only 3×10^{-6} for the MDA reaction.*

★★★ 5. **Lage JM, Leamon JH, Pejovic T, et al.** (2003) *Genome Res.* 13, 294–307. – *Demonstration of CGH by microarray using an MDA product.*

★★★ 6. Hosono S, Faruqi AF, Dean FB, *et al.* (2003) *Genome Res.* **13**, 954–964. – *Comprehensive analysis of MDA amplification bias across 47 human loci with less than sixfold bias. Report on MDA directly from a variety of clinical specimens.*

7. Sorensen KJ, Turteltaub K, Vrankovich G, Williams J & Christian AT (2004) *Anal. Biochem.* **324**, 312–314.

★★ 8. Paez JG, Lin M, Beroukhim R, *et al.* (2004) *Nucleic Acids Res.* **32**, e71. – *Thorough analysis of MDA product performance in genotyping and genome coverage.*

★★★ 9. Lasken RS & Egholm M (2003) *Trends Biotechnol.* **21**, 531–535. – *Review of MDA applications and its impact on numerous scientific fields.*

10. Luthra R & Medeiros LJ (2004) *J. Mol. Diagn.* **6**, 236–242.

11. Lovmar L, Fredriksson M, Liljedahl U, Sigurdsson S & Syvanen AC (2003) *Nucleic Acids Res.* **31**, e129.

12. Mai M, Hoyer JD & McClure RF (2004) *J. Clin. Pathol.* **57**, 637–640.

13. Esteban JA, Salas M & Blanco L (1993). *J. Biol. Chem.* **268**, 2719–2726. – *Fidelity of φ29 DNA polymerase.*

14. Tranah GJ, Lescault PJ, Hunter DJ & de Vivo I (2003) *Biotechnol. Lett.* **25**, 1031–1036.

★★ 15. Barker DL, Hansen MS, Faruqi AF, *et al.* (2004) *Genome Res.* **14**, 901–907. – *Comprehensive study of genotyping accuracy using amplified DNA.*

★★★ 16. Kornberg A & Baker TA (1991) *DNA Replication*, 2nd edn. WH Freeman & Company. – *An outstanding reference book on DNA replication and many of the enzymes used in biotechnology including DNA polymerases, reverse transcriptases, RNA polymerases, and ligases. Highly recommended for any laboratory bookshelf.*

17. Saiki R, Scharf S, Faloona F, *et al.* (1985) *Science*, **230**, 1350–1354.

18. Mullis KB & Faloona, F (1987) *Methods Enzymol.* **155**, 335–350.

19. Zhang L, Cui X, Schmitt K, Hubert R, Navidi W & Arnheim N (1992) *Proc. Natl. Acad. Sci. U. S. A.* **89**, 5847–5851.

20. Telenius H, Carter NP, Bebb CE, Nordenskjold M, Ponder BA & Tunnacliffe A (1992) *Genomics*, **13**, 718–725.

21. Cheung VG & Nelson SF (1996) *Proc. Natl. Acad. Sci. U. S. A.* **93**, 14676–14679.

22. O'Donnell ME & Kornberg A (1985) *J. Biol. Chem.* **260**, 12875–12883.

23. Lizardi PM, Huang X, Zhu Z, Bray-Ward P, Thomas DC & Ward DC (1998) *Nat. Genet.* **19**, 225–232.

★★★ 24. Raghunathan A, Ferguson HR, Bornarth CJ, Driscoll M & Lasken RS (2005) *Appl. Environ. Microbiol.* **71**, 3342–3347. – *First report on MDA from single bacterial cells.*

★★★ 25. Blanco L, Bernad A, Lazaro JM, Martin G, Garmendia C & Salas M (1989) *J. Biol. Chem.* **264**, 8935–8940. – *Pioneering work on φ29 DNA polymerase.*

26. Buckley PG, Mantripragada KK, Benetkiewicz M, *et al.* (2002) *Hum. Mol. Genet.* **11**, 3221–3229.

★★★ 27. Lasken RS, Hosono, S & Egholm, M (2004) In: *DNA Amplification: Current Technologies and Applications*, pp. 267–290. Edited by VV Demidov and NE Broude. Norwich, UK: Horizon Bioscience. – *Detailed report on the use of many types of DNA and biological specimens in MDA reactions.*

28. Harty LC, Garcia-Closas M, Rothman N, Reid YA, Tucker MA & Hartge P (2000) *Cancer Epidemiol. Biomarkers Prev.* **9**, 501–506.

29. Yan J, Feng J, Hosono S & Sommer SS (2004) *BioTechniques*, **37**, 136–143.

30. Saenger W (1984) *Principles of Nucleic Acid Structure*. Edited by CR Cantor. New York: Springer-Verlag.

31. Hughes S, Lim G, Beheshti B, *et al.* (2004) *Cytogenet. Genome Res.* **105**, 18–24.

32. Rook MS, Delach SM, Deyneko G, Worlock A & Wolfe JL (2004) *Am. J. Pathol.* **164**, 23–33.

33. Cardosa J, Molenaar L, de Menezes RX, *et al.* (2004) *Nucleic Acids Res.* **32**, e146.

CHAPTER 9

Multiple displacement amplification from single bacterial cells

Roger Lasken[1], Arumugham Raghunathan[2], Thomas Kvist[3], Thomas Ishøy[1], Peter Westermann[3], Birgitte K. Ahring[3] and Robert Boissy[1]

[1]Center for Genomic Sciences, Allegheny-Singer Research Institute, Pittsburgh, Pennsylvania, USA; [2]Qiagen Inc., 27220 Turnberry Lane, Valencia, California 91354, USA; [3]Biocentrum-DTU, The Technical University of Denmark, Denmark

1. INTRODUCTION

Multiple displacement amplification (MDA) (1–5) is capable of amplifying genomic DNA from a single bacterium (6). The efficient utilization of high-molecular-weight DNA template by the processive φ29 DNA polymerase contributes to the selective amplification of even single copies of a genome (see Chapter 8 by Lasken for details of the MDA reaction). For example, about 74% of the genome was well represented in any single MDA reaction from an individual *Escherichia coli* bacterium. The typical yield of about 24 µg of amplified genomic DNA corresponded to about a 5 billionfold amplification from the individual cell. The amplified DNA is suitable for use in species identification, genotyping, and DNA sequencing applications.

Single bacteria can be isolated by sorting with a flow cytometer, which ensures that the DNA substrate for most amplifications is from only one cell. Simpler methods of serial dilution can also be employed to obtain one cell (or a few cells) rapidly from a complex population if the cells are dominant. Also, a powerful new approach is presented in this chapter: the use of fluorescent *in situ* hybridization (FISH) combined with cell isolation by microcapillary micromanipulation (7). This strategy enables an unprecedented ability to investigate new species. MDA is also applicable to metagenomics (8, 9): the rapidly developing field of the study of complex microbial flora isolated from natural environments. It is widely accepted that only a small fraction of microbial species are known, and that a major obstacle in identifying unknown species is the lack of methods to culture them. However, MDA enables WGA and genomic sequencing directly from unculturable

Whole Genome Amplification: *Methods Express* (S. Hughes and R. Lasken, eds.)
© Scion Publishing Limited, 2005

species. As discussed below, this should complement the metagenomic shotgun approach (9) of sequencing a mixture of species present in complex microbial communities.

In this chapter, laboratory methods will be presented for single-cell isolation and amplification of genomic DNA by MDA. Novel strategies for subsequent genomic DNA sequence analysis and the design of whole-genome assembly algorithms will also be suggested. In addition to genomic DNA sequence analysis, there is a variety of other potential applications for the use of MDA-generated DNA from single cells of unculturable microbial species. These otherwise unattainable DNA products can be used to gain a better understanding of species diversity in environmental and medical specimens. In other studies, only some members of a population may be culturable, and these may be present in a variety of physiological states affecting growth rates. Also, many microbes cannot be cultured when separated from other members of their natural community that may provide metabolic precursors or other environmental requirements such as for members of biofilm communities. MDA provides a way to identify microbes without requiring the highly selective step of first culturing those microbes that are favored for growth. MDA-generated DNA has also been shown to be well suited for use in high-throughput genotyping assays (2, 10). MDA could be used to determine genetic heterogeneity within populations or species, and to study processes such as horizontal gene transfer and its role in species adaptability and evolution. The highly specific amplification from one or a few cells will also enable the development of sensitive diagnostic assays and methods to detect low levels of natural pathogens or biological weapons in the environment, or in food and water supplies.

2. METHODS AND APPROACHES

2.1. Isolation of single cells by serial dilution

Diluting cells to very low concentrations is one very simple approach to isolating them into separate reaction wells. This will be most effective for homogeneous cultures or to obtain cells of the dominant species present in communities. The cells can be diluted to an appropriate concentration and delivered to the MDA reaction wells such that the majority of wells will contain one or no cells. A predetermined average number of cells can also be delivered for protocols calling for more than one cell. A Poisson distribution can be used to determine the expected frequency of obtaining no cells, one cell, or a desired average number of cells based upon the starting titer.

The Poisson distribution can be written as follows:

$$P(k) = e^{-m}m^k/k!$$

where $P(k)$ is the probability of a well receiving k cells and m is the number of cells per chosen volume to be added to the wells. For example, if the stock of diluted cells has one cell per 10 µl and 10 µl is added to each well, then the probability of a well getting no cells = 0.37, one cell = 0.37, two cells = 0.18, three cells = 0.06, and so on.

2.2. Isolation of single cells by flow cytometry or micromanipulation

For more precise control in delivering one cell, or a pre-determined number of cells, we have used flow cytometry (6) (see *Protocol 1*). A more powerful method for the analysis of putative new species observed by microscopy involves the capture of single cells by micromanipulation (see *Protocols 5–8*).

3. RECOMMENDED PROTOCOLS

Protocol 1

Fluorescence-activated cell sorting (FACS) flow cytometry
In this example, *E. coli* was used as a test case.

Equipment and Reagents
■ Cells. *E. coli* cells were obtained from ATCC (#700926)
■ LB broth (10 g tryptone; 5 g yeast extract; 10 g NaCl; dissolved in 1 l of distilled water)
■ Phosphate-buffered saline (PBS) (Sigma)
■ FACS Vantage flow cytometer with CELLQUEST and CYTOCOUNT software (Becton Dickinson)
■ TE buffer (10 mM Tris-HCl (pH 7.5); 1 mM EDTA)

Method
1. Grow *E. coli* cells in liquid culture (LB broth) following the supplier's protocol.

2. Harvest cells in log phase and quickly cool.

3. Dilute cells in PBS and sort on a FACS Vantage flow cytometer using CELLQUEST and CYTOCOUNT software.

4. Excite cells at 488 nm and collect the emission forward scatter and fluorescence through a 530/30 band-pass filter. Analyze cells at a rate of approximately 3000 cells/s.

5. Gate the stream of FL1-positive cells to eliminate cell clusters, debris, and noise.

6. Use an automatic cell deposition unit to sort (i.e. well above the noise scatter) one, 10, 50, or 100 cells (as needed) into individual wells (containing 10 μl of TE buffer) of 96-well plates.

Protocol 2

Cell lysis[a]

Equipment and Reagents
- Denaturation solution (400 mM KOH; 10 mM EDTA)[b]
- Neutralization solution (800 mM Trizma hydrochloride (approximately pH 4; unadjusted) (Sigma)
- TE buffer (10 mM Tris-HCl (pH 7.5); 1 mM EDTA)
- Plastic tubes or microtiter plates suitable for cell lysis steps and the subsequent MDA reaction[c]

Method
1. Bring isolated cells to a volume of 26 µl with TE buffer.
2. Add 3.5 µl of denaturation solution to the cell, mix, and incubate for 5 min at 65°C.
3. Add 3.5 µl of neutralization solution and mix thoroughly by pipetting (final volume 33 µl).
4. Use in MDA reaction within 1 h.

Notes

[a]Cell lysis procedures will need to be optimized for the specific requirements of different microbial cells or cell populations.
[b]The denaturation solution can be stored for 1 week at room temperature in a tightly closed bottle to prevent neutralization.
[c]The cell should be placed in a well (or tube) suitable for carrying out the subsequent MDA reaction. To avoid the loss of any DNA that might result from a transfer step, the lysed and neutralized cell will remain in this well (or tube) throughout the procedure, with MDA reagents being added to it (see *Protocol* 3). A 96-well microtiter plate is convenient for carrying out the MDA reaction for multiple cells and for controlling the 30°C temperature of the MDA reaction in a thermocycler. Individual PCR tubes are also suitable.

When dealing with single cells, quantitative release of DNA is the ideal, as the fraction of the genomic DNA template that is available for amplification by MDA is proportional to the efficiency of its release from the cell. A number of methods have been developed for cell lysis prior to MDA, including: the use of KOH at room temperature (2, 10); the method described above using KOH at 65°C (6); and 95°C treatment under neutral conditions (1). Freeze–thaw treatment has been suggested for the lysis of archaebacteria. We have tested five to ten cycles of quick freezing by dipping the bottom of the tube or microtiter plate in a dry ice/ethanol bath for 1 min and thawing (which we accelerated by dipping the tube in warm water). However, controlled experiments were not carried out to determine the precise number of freeze–thaw cycles required, and the empirical optimization of this step is recommended. Some cells – especially spores – are expected to be more difficult to lyse than *E. coli*. Sonication methods or treatment with an electric pulse are currently under investigation by a number of laboratories. Regardless of the cell lysis protocol used, it must be remembered that, although quantitative release of DNA is the ideal, MDA is inefficient for

Protocol 3

MDA reaction[a]

Equipment and Reagents
■ REPLI-g Kit (4× REPLI-g buffer containing exonuclease-resistant, phosphorothioate-modified, random hexamer oligonucleotide primers; REPLI-g DNA polymerase (φ29 DNA polymerase); nuclease-free water; Qiagen)

Method
1. Prepare a master mix of 41 µl of nuclease-free water, 25 µl of 4× REPLI-g buffer and 1 µl of REPLI-g DNA polymerase[a] (final volume here 67 µl) for each reaction.

2. To the 33 µl of solution containing a lysed and neutralized cell from *Protocol 2*, add 67 µl (from step 1) of master mix (final reaction volume 100 µl) and mix well by pipetting.

3. Incubate at 30°C on a thermocycler for 16 h or overnight[b].

4. Terminate the reaction by raising the temperature to 65°C for 3 min to inactivate the enzyme.

5. Store amplified DNA at 4°C if it is to be used immediately or at –20°C for long-term storage.

Notes
[a]Protocols for MDA WGA vary according to manufacturers' instructions (see section 2.3 and *Protocol 1* in Chapter 8). In the example here for single *E. coli* cells (6), the REPLI-g Kit (originally distributed by Molecular Staging and now distributed by Qiagen) was used.
[b]A thermocycler can be conveniently programmed to carry out the 30, 65, and 4°C steps in this protocol.

---

highly fragmented DNA (see Chapter 8, section 3.3). Therefore, cells should be lysed in the gentlest manner possible. Furthermore, DNA as large as bacterial genomes will be unavoidably fragmented when it is released from cells, and sequences adjoining break points in haploid genomes may not be favorably represented in the amplified DNA.

3.1. DNA sequencing with amplified DNA

Genomic DNA libraries have been constructed and sequenced from MDA reaction products (3). We have also investigated the use of PCR to enrich target sequences from the amplified DNA for use in cycle sequencing. This approach can be used to obtain initial DNA sequence data, interrogate specific genes of interest, and as a rapid method for the genetic analysis of complex populations. A PCR amplification step also ensures that sufficient DNA is available from sequences that may be underrepresented in an MDA WGA reaction.

A three-step protocol was tested that used:

1. MDA to amplify the whole genome from a single *E. coli* cell.
2. PCR to enrich a specific target region for DNA sequencing.
3. Cycle sequencing.

This approach can be used when the microbe's genomic DNA sequence is available for primer design, as was the case for *E. coli*, and could enable some new strategies for investigating ecosystems at the level of their individual member species. For example, research into the genetic diversity of populations, mutation frequencies, or horizontal gene transfer could be carried out in this manner using either cycle sequencing, or DNA sequencing-based single nucleotide polymorphism (SNP) and mutation assays.

Prior to investing resources in DNA sequencing, the presence of specific target sequences in MDA reactions can be tested by resolving aliquots of PCR products by agarose gel electrophoresis. Other aliquots from the PCR could then be processed for cycle sequencing if they have the expected band. Alternatively, cycle sequencing could be carried out on all PCR products, and analysis of the sequences obtained could then be used to evaluate and optimize the efficiency of the cell isolation, cell lysis, and MDA steps (see *Protocols 1–3*).

In this experiment, the presence of the target sequences in the MDA WGA was first verified by pre-screening ten targets (6) by TaqMan PCR to determine those MDA reactions suitable for use in sequencing. Next, to test the fidelity of sequencing, a total of 50 PCRs were carried out using ten PCR primer sets used on each of five replicate MDA reactions, each carried out from a single *E. coli* cell. Each PCR received 2.0 µl (approx. 1.6 µg of DNA) from a 100 µl MDA reaction. The PCR products were processed with a QIAquick PCR Purification Kit (Qiagen), and 2.5 µl of this was used as a template for cycle sequencing. All ten targets were successfully sequenced (see *Table 1*) for all five replicate MDA reactions with 100% concordance to the known sequences. Therefore, highly specific genomic sequence from one cell is conserved in the MDA reaction, and the MDA product is a suitable template for robust PCR amplification. For controls lacking the single cell in the MDA reaction, no PCR product was obtained.

Table 1. Sequencing from PCR products derived from MDA WGA from a single cell

In controls (right column), the flow-sorted single cell was omitted from the MDA reaction and PCR and cycle sequencing were carried out as normal. The PCR primers used have been published (6). Reprinted from (6) with permission of *Applied and Environmental Microbiology*.

Locus name	Amplicon size (nt)	Identity with GenBank (%)	No cell present in MDA
CadA	211	100	0
EutC	191	100	0
Exo	154	100	0
ExuR	179	100	0
GlyS	186	100	0
HolA	71	100	0
Nth	220	100	0
OmpA	90	100	0
PcnB	158	100	0
TopA	191	100	0

Protocol 4

PCR and cycle sequencing of MDA reaction products[a]

Equipment and Reagents
- DNA amplified from a single cell (see *Protocols 1–3*)
- Appropriate PCR primers and cycle sequencing primers
- Reagents for carrying out PCR (e.g. from Invitrogen and Applied Biosystems) and cycle sequencing (e.g. from Amersham Biosciences).
- QIAquick PCR Purification Kit (Qiagen)

Method
1. Add 2 μl (approx. 1.6–2 μg of DNA) of the MDA reaction to a standard PCR[b].

2. Purify PCR products using the QIAquick PCR Purification Kit, following the manufacturer's instructions[c].

3. Carry out cycle sequencing with 2.5 μl of the cleaned PCR product[c], following the manufacturer's instructions.

Notes

[a]There are several advantages of using MDA for DNA sample preparation for use in PCR detection and cycle sequencing. PCR can be capable of amplifying DNA from a single cell without the need for the MDA step, but considerable optimization is often needed. Moreover, when the target DNA is present in a highly complex background of sequences from other species the challenge is even greater. Use of MDA prior to PCR was recently shown to greatly improve detection of the bacteria *Wolbachia* from total DNA extracted from infected mites (11) and single fungal spores (12). MDA could be useful in detecting the presence of microbes in many environmental samples, medical specimens, and food and water supplies, and in early warning for emerging diseases and biological weapons.

[b]Hot-start PCR protocols are preferred due to the complexity of the amplified genomic DNA target (for example, use Platinum *Taq* DNA polymerase from Invitrogen). We tested a protocol with 40 cycles of PCR and amplicons of 100–300 nt, and used 2 μl of the MDA reaction as template. Other conditions were not tested.

[c]It is essential to remove PCR primers prior to cycle sequencing. Cycle sequencing reactions were carried out at the HHMI/Keck Biotechnology Resource Laboratory, Yale University, USA. Commercial kits are also available to carry out cycle sequencing (e.g. from Amersham Biosciences). One of the PCR primers used in step 1 can be employed as the cycle sequencing primer; alternatively, a nested primer internal to the PCR primers is often used to improve specificity.

There are several advantages to using an MDA reaction prior to PCR analysis or cycle sequencing. Even when PCR can be optimized for use with a single cell, only one or a few multiplexed targets can be obtained since the cell is sacrificed in the process. MDA product can be used as template for any number of PCR targets without the challenge of multiplexing, and the unused amplified DNA is valuable as an archive for future use since the cell is sacrificed. MDA also provides unlimited template for PCR, which fails more frequently when template is limiting. MDA allows addition of nanograms or more of the template to the PCR compared with femtograms of template if MDA is not used (e.g. about 5 fg of DNA in a single *E. coli* cell). The abundant DNA generated in an MDA reaction can also be used to carry out replicate downstream assays to confirm results.

3.2. Characteristics of DNA amplified by MDA from single cells

3.2.1. Specific amplification of substrate genomic DNA

MDA reactions were carried out with different numbers of *E. coli* cells (100, 50, 10 or one) to test the specificity of the reaction and its ability to amplify low copy numbers of genomes (see *Fig. 1* in color section). First, we confirmed the ability to isolate single cells successfully by FACS flow cytometry (see *Fig. 1a* in color section). These cells were plated in a microplate, and wells contained either one or no cells (see *Fig. 1b* in color section).

Having verified the ability to use FACS flow cytometry efficiently, we next collected single cells for testing in MDA reactions. In other wells, FACS flow cytometry was used to deliver an average of 100, 50, or 10 cells. The cells were lysed and their DNA was used in MDA reactions. TaqMan PCR assays were then used to quantify the presence of several *E. coli*-specific loci in the MDA reaction products. A locus representation value of 100% indicates that the locus is present in the same copy number in the amplified DNA as in the starting template. The average loci representation values tended to drop for amplifications from fewer cells (see *Fig. 1c* in color section). The two loci tested for MDA reactions containing a single cell were only represented at an average of about 25% compared with unamplified genomic DNA. Clearly, MDA from a single cell is challenging and will require optimization for different microbes. Nevertheless, even at an average representation of 0.25 copies per MDA-generated genome equivalent, this represents an enormous amplification of over a billionfold for these loci. In a more complete analysis (see below), ten different loci were represented with an overall average of 30%. This corresponds to about a 5×10^9-fold amplification of DNA from the 5 fg of DNA present in one *E. coli* cell.

3.2.2. Amplification bias

In order to investigate the quality and coverage of the amplified DNA, TaqMan assays were used to evaluate the copy number of ten different loci in the *E. coli* genome. When non-limiting amounts of template DNA are used, MDA is capable of amplifying the majority of a genome with less than a sixfold difference in the yield of different loci (10). MDA from single cells has a much wider range of amplification bias (6). A comprehensive study was carried out measuring the representation of ten loci in 84 amplifications each from a single *E. coli* cell (see *Table 2*). There was about a 58% chance that a given sequence was represented in the amplified DNA at >0.1% of its copy number in the starting *E. coli* genomic DNA template. The efficiency of MDA was probably considerably higher than 58%, since not all reactions successfully received a cell from flow cytometry (e.g. see *Fig. 1b* in color section). In this experiment, 18 of the 84 MDA reactions failed to show amplification of any *E. coli* sequences and were assumed not to have received a cell.

The flow cytometry setup was designed to capture either a single cell or no cell in most wells and avoid the accidental collection of multiple cells (see *Fig. 1a* and *b* in color section). For the 66 wells that were positive for *E. coli* DNA amplification, a given locus was represented approximately 74% of the time (489 positive detections out of the 660 opportunities from 66 MDA reactions and ten test loci). Loci representation values (see *Table 2*) ranged from 0.1% (the lower limit of detection in this quantitative PCR experimental design) to 1211%. Stochastic effects of starting with a single genomic copy probably result in greater variability in yields of different sequences. However, even 0.1% efficiency corresponds to a 10^6–10^7-fold DNA amplification from the single cell and will enable many kinds of detection, diagnostic, and sequencing applications that are based on specific primers or probes such as in PCR and array hybridization. Also, many of the dropout loci not detected in this assay were presumably present with lower levels of amplification of $\leq 10^5$-fold and may support analysis in other optimized PCR-based assays.

3.2.3. Pooling strategies to average out amplification bias

The loss of specific loci (i.e. their inability to be amplified) during MDA from single bacteria is fairly random, having occurred on average to about the same extent for all ten loci tested (see *Fig. 2*). Each bar represents, for the indicated locus, the average result obtained for the 84 MDA reactions. The locus representation averaged across all ten loci tested was about 30% (see *Fig. 2*, far right bar). While different sequences may vary somewhat in their efficiency of amplification, all were well represented in at least some of the amplifications. No locus had an average representation higher than 54% or lower than 15% for the 84 MDA reactions. The random nature of the sequence dropout means that DNA or cell-pooling strategies could be employed to increase the percentage of the genome that can be represented, for example, for DNA sequencing applications. Separate MDA reactions could be carried out, each receiving a different individual cell. These cells can be selected on the basis of belonging to the same species by morphological observation and also by a new method of FISH labeling and cell isolation by micromanipulation (see section 3.3 below).

A pooling strategy may be particularly important for the construction of genomic DNA libraries. In the example here (see *Table 2*), all ten loci could be represented to a similar extent by pooling multiple MDA reaction products (each obtained from a single cell), or carrying out one MDA reaction with multiple cells of the same species. In previous work, a genomic DNA library was constructed and sequenced using DNA obtained by MDA of about 1000 cells of the bacterium *Xylella fastidiosa* (3). Portions of the library were sequenced to a coverage depth of approximately sevenfold, and contigs could be assembled from these data using PHRAP. One should expect MDA reactions from single cells to exhibit greater amplification bias than those reactions using non-limiting amounts of DNA; however, pooling strategies can be expected to ameliorate the underrepresentation of specific loci in the resulting libraries.

Table 2. Representation of ten *E. coli* loci in MDA-generated DNA from single cells

A total of 84 MDA reactions were carried out, each using a single flow-sorted *E. coli* cell to provide the DNA template. TaqMan PCR assays were used to quantify the presence of each locus and its relative representation in the MDA-generated reaction products (6). A locus representation of 100% indicates that the sequence is present in the same copy number per genome equivalent in the amplified DNA as in the unamplified genomic DNA template. Eighteen reactions show no amplification and were assumed to have failed to receive a cell from flow cytometry. For reactions that did receive a single cell, loci were detected (above a 0.1% representation threshold) an average of 74% of the time. Reprinted from (6) with permission of *Applied and Environmental Microbiology*.

MDA #	Locus ID and locus representation (copy number in MDA product per genome equivalent, %)									
	Exo	*OmpA*	*GlyS*	*TopA*	*HolA*	*CadA*	*ExuR*	*PcnB*	*EutC*	*Nth*
1	1.4	0.0	0.1	0.0	14.5	0.3	0.0	0.2	0.0	2.3
2	0.3	0.0	6.5	1.7	0.6	3.6	7.6	0.0	0.9	0.4
3	139.0	95.3	0.0	0.0	4.0	0.5	0.0	0.0	10.7	0.5
4	0.0	52.4	0.3	0.2	0.4	0.0	0.0	0.0	0.7	24.7
5	62.1	102.1	0.0	0.3	0.0	20.8	15.4	0.0	0.1	0.1
6	130.2	0.6	0.0	0.1	4.6	0.0	0.2	12.2	2.1	1.1
7	45.4	0.0	0.0	0.0	0.0	2.1	0.1	0.0	0.0	0.0
8	0.0	11.2	28.6	0.0	2.8	0.0	0.1	0.0	0.2	0.1
9	0.7	77.1	20.7	230.5	50.0	8.8	12.3	1.0	12.6	74.2
10	0.0	0.0	0.0	0.0	0.0	0.0	0.0	0.0	0.0	0.0
11	55.3	214.8	29.4	56.8	17.4	20.5	8.5	35.7	52.2	58.6
12	119.7	90.0	31.9	55.5	15.4	22.2	23.4	14.3	1.6	65.1
13	41.5	260.9	17.9	4.2	0.1	0.0	5.5	0.0	39.0	13.8
14	0.0	0.0	0.0	0.0	0.0	0.0	0.0	0.0	0.0	0.0
15	0.0	0.0	0.0	0.0	0.0	0.0	0.0	0.0	0.0	0.0
16	2.4	0.9	0.0	9.4	737.0	0.0	140.7	1.2	19.6	0.0
17	100.6	0.0	0.0	0.4	0.0	2.8	0.2	0.0	0.0	0.0
18	93.9	6.3	0.0	0.0	0.0	169.0	6.0	0.4	0.0	0.0
19	0.1	0.0	0.0	0.0	0.4	0.0	0.1	0.0	0.0	0.0
20	4.5	0.0	54.6	59.1	1.3	67.2	127.7	2.4	31.5	2.6
21	226.4	1.0	0.0	67.1	100.0	0.0	28.7	14.2	74.6	0.6
22	8.4	178.7	0.9	146.4	293.5	0.6	1.8	8.7	9.6	2.3
23	0.0	0.5	1.6	214.0	1.9	26.5	0.1	14.9	0.4	0.0
24	14.8	3.0	136.0	12.7	11.7	18.9	2.7	0.9	0.4	47.5
25	7.7	63.2	167.8	141.2	14.2	0.4	17.3	5.2	0.0	44.6
26	1.3	7.2	0.0	0.0	0.0	26.9	14.9	257.3	0.0	1.8
27	60.9	23.5	24.7	39.7	16.6	49.4	24.1	39.8	16.2	4.2
28	0.0	0.0	53.7	0.0	0.0	11.4	0.2	0.0	1.3	0.0
29	84.5	24.1	33.3	1.9	302.0	0.6	0.0	1.6	0.9	0.0
30	19.5	0.0	0.0	0.6	0.0	26.8	76.7	11.3	0.0	266.1
31	49.2	1.2	58.6	14.3	71.2	18.8	135.4	6.4	104.5	10.8
32	0.0	8.7	143.2	0.0	0.0	5.6	4.7	0.0	6.0	0.1
33	0.0	0.0	0.0	0.0	0.0	0.0	0.0	0.0	0.0	0.0
34	21.1	0.0	195.9	0.4	20.1	0.0	84.4	6.1	0.1	5.1
35	0.0	0.0	0.0	0.0	0.0	0.0	0.0	0.0	0.0	0.0
36	48.2	82.7	0.9	109.1	50.9	1.9	0.3	2.0	4.4	265.6
37	0.0	0.0	0.0	0.1	0.0	0.0	569.3	0.0	0.0	1.6
38	32.3	56.2	6.9	4.4	569.4	27.6	150.5	0.6	4.1	0.8

	Exo	OmpA	GlyS	TopA	HolA	CadA	ExuR	PcnB	EutC	Nth
39	0.0	0.0	0.0	0.0	0.0	0.0	0.0	0.0	0.0	0.0
40	16.4	79.6	86.5	25.8	39.5	49.2	0.0	16.9	30.6	20.8
41	29.4	22.3	43.9	12.8	4.1	0.7	6.8	0.4	0.1	0.7
42	0.0	0.0	0.0	0.0	0.0	0.0	0.0	0.0	12.6	128.5
43	98.1	25.3	0.0	0.1	25.0	30.3	7.2	21.1	0.6	10.4
44	0.1	22.4	82.4	88.6	1.6	0.7	0.0	0.7	55.2	177.0
45	0.0	0.0	0.0	0.0	0.0	0.0	0.0	0.0	0.0	0.0
46	0.0	0.0	37.9	0.1	138.5	0.0	445.9	0.0	0.1	0.0
47	0.0	24.0	0.0	39.2	5.8	0.2	0.5	0.0	0.5	1.0
48	1.6	0.3	28.7	1.1	305.2	87.3	0.0	11.1	89.3	38.8
49	11.3	0.2	50.1	40.9	14.0	61.4	3.8	93.4	0.0	260.8
50	0.2	17.7	8.5	14.6	0.1	0.0	21.9	15.4	4.1	0.3
51	90.1	0.1	0.0	0.0	0.0	207.2	0.0	0.0	0.0	0.0
52	16.0	6.0	7.5	0.5	107.9	11.2	25.6	3.3	88.2	272.7
53	33.2	47.6	20.3	104.9	91.1	0.3	14.7	32.4	7.4	252.9
54	10.4	24.8	9.7	5.9	0.0	0.9	112.5	0.1	0.6	59.2
55	0.0	0.0	13.6	0.1	0.0	0.0	0.0	126.8	0.0	0.0
56	15.8	512.6	7.4	20.1	0.0	19.2	299.0	0.0	1.0	946.5
57	0.0	0.0	0.0	0.0	0.0	0.0	0.0	0.0	0.0	0.0
58	0.0	0.0	0.0	0.0	0.0	0.0	0.0	0.0	0.0	0.0
59	0.0	0.0	0.4	25.2	0.0	0.0	0.2	0.0	7.1	0.0
60	25.2	97.1	12.0	86.2	184.3	4.1	51.1	21.9	1.6	0.8
61	67.6	5.3	0.0	0.0	276.9	23.0	0.0	0.5	11.4	86.1
62	0.0	0.0	0.0	0.0	0.0	0.0	0.0	0.0	0.0	0.0
63	0.0	0.0	0.0	0.0	0.0	0.0	0.0	0.0	0.0	0.0
64	3.2	0.0	83.9	9.4	1.2	137.0	3.4	14.2	2.1	19.9
65	5.2	219.8	49.0	18.2	82.7	27.7	8.7	173.2	144.1	0.0
66	0.0	0.0	0.0	0.0	0.0	0.0	0.0	0.0	0.0	0.0
67	3.0	6.9	0.6	6.0	0.0	0.0	0.9	0.8	736.4	1.1
68	73.3	304.5	12.8	41.9	21.4	0.0	28.9	0.8	16.8	65.7
69	0.0	0.0	0.0	0.0	0.0	0.0	0.0	0.0	0.0	0.0
70	0.0	0.0	0.0	0.0	0.0	0.0	0.0	0.0	0.0	0.0
71	0.6	164.8	2.8	0.0	41.6	2.2	9.9	2.6	0.6	0.0
72	0.1	1211.1	68.9	3.6	103.0	2.7	35.5	111.2	0.0	0.1
73	0.1	67.5	50.9	53.5	24.2	2.3	0.4	18.6	0.0	236.5
74	0.0	11.3	0.0	68.1	52.5	19.7	0.0	11.0	0.0	187.7
75	0.0	0.0	0.0	0.0	0.0	0.0	0.0	0.0	0.0	0.0
76	0.0	0.0	0.0	0.0	0.0	0.0	0.0	0.0	0.0	0.0
77	175.6	6.2	24.2	26.1	233.7	11.7	37.9	7.5	22.1	23.9
78	209.0	0.0	158.9	0.0	0.0	0.0	0.0	0.0	4.4	0.0
79	14.5	8.4	0.0	90.0	0.0	0.0	22.0	3.0	0.4	82.8
80	0.0	0.0	0.0	0.0	0.0	0.0	0.0	0.0	0.0	0.0
81	0.0	0.0	0.0	0.0	0.0	0.0	0.0	0.0	0.0	0.0
82	0.1	141.5	0.1	463.4	0.0	74.1	0.4	0.0	0.0	0.1
83	0.0	152.8	0.0	0.0	0.0	8.0	0.2	0.0	0.0	0.0
84	86.3	0.0	0.5	4.0	29.4	0.0	98.6	175.3	117.2	0.4
Average	28.1	54.1	22.3	28.8	48.6	15.6	32.1	15.5	20.8	44.9

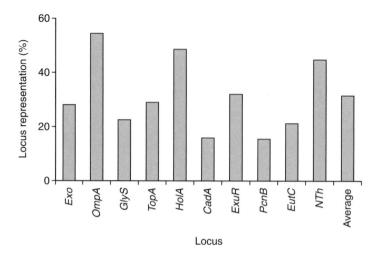

Figure 2. WGA from single *E. coli* cells.
Single cells were sorted into 84 of the 96-wells in a microtiter plate using a FACSCalibur, lysed and subjected to WGA by MDA and TaqMan quantitative PCR analysis of ten loci (see *Table 2*). Average locus representation (*n*=84) is plotted on the *y*-axis for ten individual loci in the *E. coli* genome. The average of all ten loci of about 30% is represented in the last bar. Reprinted from (6) with permission of *Applied and Environmental Microbiology.*

3.2.4. Effect of GC content on loci representation

We are not aware of any reports that MDA underrepresents sequences with a high GC content. The strong strand-displacing activity of φ29 DNA polymerase (see Chapter 8, section 1.2) appears to overcome any inhibitory effect that might be attributable to templates with higher melting temperatures. This was the case (6) for MDA using DNA from *Myxococcus xanthus*, a bacterium whose genome is approximately 68% GC. A 1 kb region – also having about 68% GC content – within the *C*-methyltransferase gene efficiently yielded amplified sequence based on TaqMan quantitative PCR (see *Fig. 3*) of MDA-generated genomic DNA.

3.2.5. Use of MDA for discovery and genomic sequencing of new species

The ability to amplify genomic DNA from a single bacterial cell should enable new research strategies for studying unculturable species. It should also be applicable to metagenomic research, the study of genomes of multi-species microbial communities in natural environments (8, 9). Complex field samples may contain DNA from dozens to thousands of different microbial species, and hence are challenging to analyze. An alternative to the MDA-mediated WGA from single cells described here has been to sequence a mixture of the entire DNA present in an environmental sample. In such a 'metagenomic shotgun' strategy, the DNA is sequenced to a sufficient depth to allow assembly of sequences for individual

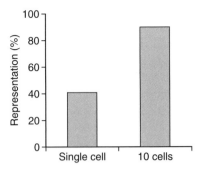

Figure 3. WGA from one or ten *Myxococcus xanthus* cells.
Amplification of single *M. xanthus* cells yielded amplified sequence based on TaqMan PCR for the *C*-methyltransferase gene (*M12-ta* locus). Five of 44 MDA reactions carried out on single flow-sorted cells were successful and had a minimum of 13% and a maximum of 74% representation of the locus. Reprinted from (6) with permission of *Applied and Environmental Microbiology.*

species (9, 13). The metagenomic shotgun method and the analysis of MDA products from single cells should prove to be complementary approaches to the highly difficult 'community genomics' analysis of complex microbial ecosystems. For example, single-cell analysis may contribute to assigning sequence-read data to individual species, confirming the assembly of contigs, and identifying low-abundance species. MDA may also be helpful as the first step in a metagenomic shotgun approach when the amount of DNA in the field sample is limiting, e.g. during the collection of air samples for the detection of natural pathogens or biological weapons. The fidelity, isothermal nature, and overall robustness of MDA recommend its engineering into microfluidic sensors (i.e. together with pathogen detection arrays and wireless data transmission).

3.3. Isolation of single cells by micromanipulation and genomic DNA amplification by MDA

A new approach to discovery of microbial species has come from the use of MDA with individual cells visualized by microscopy and captured by micromanipulation. This is providing an unprecedented ability to investigate unculturable species. Several key methodologies can be used in combination.

1. Observation of the morphology and characteristics of single cells from mixed cultures or field samples.
2. Labeling of cells with FISH probes specific for taxonomic groups of interest.
3. Isolation of the cells by micromanipulation.
4. Amplification of their genomic DNA with MDA.
5. DNA sequencing or detection of genes of interest by PCR or microarray methods.

3.3.1. Background

Traditionally, the major steps forward in the history of microbiology have been associated with the development of new and more powerful microscopes and microscopic techniques. This tradition has not always been well integrated with molecular biology and molecular biological techniques. Several recent microscopic techniques such as confocal laser microscopy and atomic force microscopy have, however, demonstrated that advanced microscopy and molecular biology might contribute synergistically to major breakthroughs in microbiology.

Since Robert Koch, Walter Hesse, and Richard Petri in the late 19th century developed the necessary techniques for isolation of microorganisms on solid agar surfaces, these principles have formed an obligate cornerstone in the establishment of axenic cultures of microorganisms. The basic principle has even been maintained for isolation of anaerobes (agar surface in a roll tube) and for extremophiles (agar exchanged with Gelrite or silica gel).

A fundamental problem associated with this technique has, however, emerged following the introduction of molecular analyses of environmental microbial populations. Numerous publications have documented that as little as 1% of the indigenous prokaryotic organisms found in natural environments are known, and probably only a fraction of this diversity can be accessed using traditional plating techniques (14). Some of the reasons for this are lack of knowledge of the specific growth conditions, specific nutrients, and obligate coculture requirements. Also, growth on a solid surface directly exposed to the atmosphere puts a very strong selective pressure on single cells supposed to develop into visible colonies.

These problems have been the driving force behind the development of techniques based on micromanipulation for isolation of single bacterial and eukaryotic cells throughout the last century (15–17). The main obstacle of these early systems was the low magnification, which was insufficient for manipulation of single prokaryotic cells. Also common to these techniques was the use of microneedles or microcapillaries for the separation of cells. Continuous development of modern microscopes in combination with the development of the modern microscopic *in vitro* fertilization techniques has further aided the manipulation of single cells. Two principles have dominated the development of micromanipulation techniques. One uses a focused laser beam to capture and move the cell of interest from a population to a compartment from where it can be transferred to a growth medium (18). The other approach is based on microinjectors in combination with the precision of a servo-powered micromanipulator, consequently enabling easy handling of a single prokaryotic cell. Fröhlich & König (19) applied such a system using pre-fabricated Bacto tips to isolate and successfully cultivate single cells from different laboratory cultures and natural environments.

Genetic knowledge about the large fraction of uncultivated microorganisms is scarce and is generally limited to the 16S rRNA genes that have been extracted from natural environments and cloned for phylogenetic analyses. Characteristics such as basic and specialized metabolism, cell morphology, and DNA–DNA homology to related species and other characteristics cannot, for obvious reasons, be determined from a gene encoding ribosomal RNA, and further insight into the function and potentials of the unculturable microorganisms has until now been impossible.

A limited insight into the world of unknown microbes has recently been achieved using FISH, by which unknown organisms can be visualized by applying specific fluorescent probes targeting 16S rRNA in corresponding cells. This technique is now widely used in environmental studies. By applying MDA and genomic sequencing to single microbial cells, labeled by FISH and isolated by micromanipulation, a quantum leap is gained in accessing information about unculturable microorganisms.

3.3.2. Micromanipulation and isolation of single cells

The micromanipulator system consists of a Leica DMIRBE inverted microscope (Leica Microsystems A/S) equipped with a MultiControl 2000 micromanipulator (ITK Lahnau) and an Eppendorf Transjector 5246 microinjector (Eppendorf AG). All the components of the system are operated from a computer system to eliminate vibration during the isolation procedure. The microscope is equipped with a mercury lamp and filter set for epifluorescence microscopy, and for documentation a Kappa DX20H camera with the computer modules CONTROL and METREO version 2.7 (Kappa ImageBase) is installed. The isolation is carried out in a chamber made of a 3 mm thick microscope slide with a 12 mm diameter hole in the center when dealing with living cells, or on a conventional slide for FISH-probed cells (see *Fig. 4a* and *b*). The bottom of the well is sealed with a cover slide by applying Vaseline to the edges. This construction results in a chamber ideal for manipulation of cells in suspension. The sample chamber is sterilized in UV light prior to the isolation experiments.

3.3.3. Microcapillaries for single-cell isolation

Microcapillaries can be obtained from several suppliers (e.g. Eppendorf). Alternatively, they can be produced in the laboratory. The microcapillaries used for isolation of single prokaryotic cells here were made from glass capillaries with an outer diameter of 1 mm. The equipment is based on a power supply (Labor Netzgerat EA-3020S) connected to a coiled heating filament. By changing the current through the filament, the coil temperature is changed, controlling the melting of the glass tube. The ends of the capillary are secured in two clamps, both being coupled to a micrometer screw. This allows very precise pulling of the capillary into a microcapillary. By viewing the area around the heating filament through a 40× magnifier (Leica M3Z Heerbrugg), it is possible to obtain a capillary with an inner diameter in the range of approximately 2–10 μm, which is suitable for handling single prokaryotic cells (see *Fig. 5*). The microcapillaries are subsequently sterilized by UV radiation for 20 min prior to use.

(a)

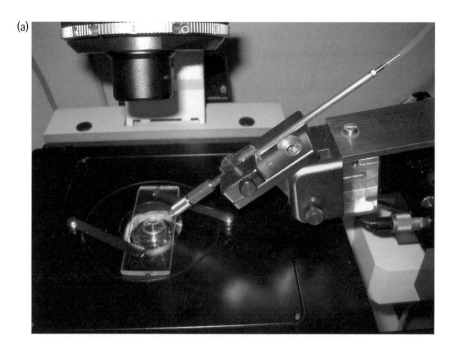

(b)

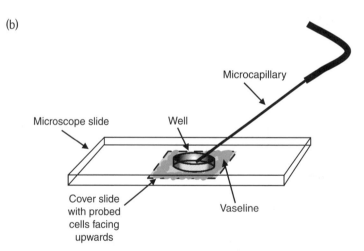

Figure 4. Selection of a cell in the micromanipulator setup (*a*) and outline of the microscopic slide (*b*).
The cover slide with probed cells is mounted underneath the slide with the probed cells facing upwards.

3.3.4. Nycodenz whole-cell separation

To separate microorganisms from the inorganic fraction of environmental samples, the sample can be subjected to Nycodenz whole-cell separation (20). The technique is based on a 1.3 g/ml solution of Nycodenz ((5-(*N*-(2,3-dihydroxypropyl)

(a)

(b)

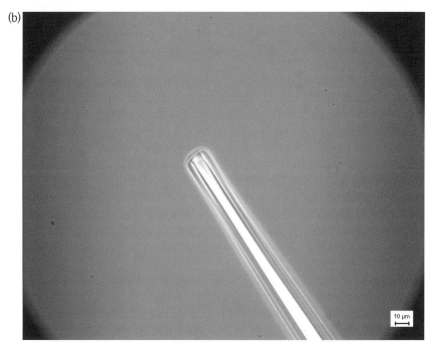

Figure 5. Production of a microcapillary.
(*a*) A capillary in front of the heating filament. The insert shows a capillary in front of the heating filament, as seen through the magnifying equipment during the pulling of a microcapillary. (*b*) A microcapillary with an inner diameter of 7 μm viewed with a 40× objective. Bar, 10 μm.

acetamido)-2,4,6-triiodo-N,N'-bis(2,3-dihydroxypropyl) isophthalamide). The Nycodenz solution has a density higher than that of the microorganisms, and the microorganisms will consequently form a layer on top of the dense liquid (see *Fig. 6*). In contrast to the microorganisms, inorganic substances will, due to their higher density, be centrifuged through the liquid, and these particles will as a result be separated from the organic fraction of the mixture and form a pellet at the bottom of the tube.

3.4. Example of single-cell isolation and MDA

The example shown consists of an agricultural soil sample where cells and bulk soil were separated by Nycodenz gradient centrifugation. The cells were probed with the FISH probe Arch 915 (5'-Cy3-GTGCTCCCCCGCCAATTCCT-3') (21), targeting mainly Archaea, but also some Gram-positive bacteria. Following detection and selection of a single fluorescent cell, all other cells were removed from the area around the cell of interest (see *Fig. 7* in color section), by means of the micromanipulator. After ensuring that the only remaining cell left in the area was the FISH-probed cell, the cell was captured with the micromanipulator and injected into TE buffer. Five freeze–thaw cycles were applied to open the cell, and a 50 μl MDA reaction (see *Protocol 8*) was used to amplify the genome of the isolated cell.

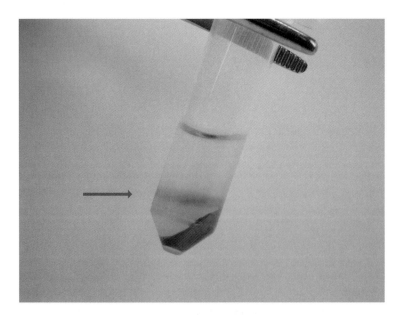

Figure 6. Nycodenz whole-cell extraction from soil.
The picture shows Nycodenz extraction after the centrifugation step (see *Protocol 5*, steps 5 and 6). The cells from the environment will be trapped in the interface between the two layers (arrow) and the inorganic fraction will form a pellet at the bottom of the tube.

PCR to amplify 16S rRNA genes followed by cloning and sequencing of the products was subsequently carried out. The phylogenetic relationship of the sequences retrieved from the FISH-probed cell is shown in *Fig. 7* (in color section) and the closest known relatives from GenBank are shown in *Fig. 8*. The sequences Klon1-H2 and Klon 3-H3 in the phylogenetic tree represent two different copies of 16 S rRNA genes retrieved from the one cell shown in *Fig. 7* (in color section). Many members of the *Bacillus* cluster are known to possess several operons encoding 16S rRNA. In molecular ecological studies based on direct DNA extraction from the environment, the possibility of assigning several 16S rRNA gene sequences to one species is not present. The example shown demonstrated that the evaluation of species diversity from the diversity of different 16S rRNA gene clones based on phylogenetic trees drawn from directly extracted DNA can be problematic. However, the combination of molecular and microscopic analyses outlined in this chapter can be combined with extended genomic sequencing and will contribute new insight with respect to the large number of unknown microorganisms found in nature.

Analyses of new microbial isolates have so far been based upon experiments with axenic cultures under laboratory conditions. Often only a tiny fraction of the metabolic capabilities of the organism is exposed in these experiments, which most frequently are made with the purpose of classifying the organism and delimiting it phenotypically from close relatives. Although the enigma of accessing the unculturable microbial majority of most ecosystems might be

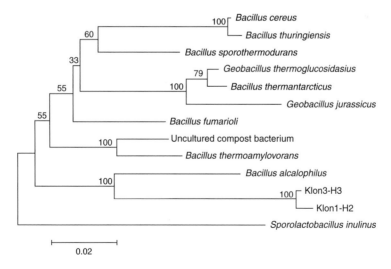

Figure 8. Phylogenetic tree of cloned 16S rRNA genes and their closest relatives.
The figure shows the phylogenetic relationship of the retrieved sequences and the closest matching sequences from the GenBank database. The tree is constructed as a Jukes & Cantor neighbor-joining tree. Numbers above branchings are bootstrap values. Values below 40 are not shown. The scale indicates the number of substitutions per nucleotide. Klon3-H3 and Klon1-H2 represent the two sequenced clones of this study.

Protocol 5

Whole-cell separation and fixation

Equipment and Reagents
- Nycodenz (1.3 g/ml; Sigma)
- 0.9% NaCl
- 1:1 Ethanol/PBS (50 ml 100% ethanol and 50 ml PBS (Sigma))

Method
1. Transfer 300 µl of Nycodenz to a 2 ml Eppendorf tube and place the tube at 4°C for 5 min.

2. Mix 1 g of soil with 5 ml of 0.9% NaCl and vortex the sample for 1 min.

3. Leave the suspension at 4°C for 10 min.

4. Vortex for 10 s.

5. Place 600 µl of the slurry on top of the Nycodenz.

6. Centrifuge the Eppendorf tube for 10 min at 10 000 *g*.

7. Transfer 600 µl of the middle (and top) phase to a fresh tube (see *Fig. 6*, arrow).

8. Centrifuge the Eppendorf tube at 10 000 *g* at 4°C for 2 min and discard the supernatant.

9. Add 1 ml of 0.9% NaCl and mix well.

10. Centrifuge the Eppendorf tube at 10 000 *g* at 4°C for 2 min and discard the supernatant.

11. Resuspend the pellet in 500 µl of 1:1 ethanol/PBS[a].

12. Place the sample at –20°C for 8 h.

Notes
[a]It may be necessary to alter the volume of 1:1 ethanol/PBS that the pellet is resuspended in to obtain a good spread of chromosomes for *Protocol 6.*

partially solved by laborious developments of new cultivation techniques, a significant portion of the indigenous microorganisms will probably remain unculturable. The rapid progress and decreasing costs and labor of whole genome sequencing and analysis techniques in combination with the new techniques outlined in this chapter might challenge the current paradigm of necessary isolation and cultivation procedures to characterize new microorganisms.

Protocol 6

FISH

Equipment and Reagents
- 2.5 × 7.5 cm glass cover slides
- Formamide (Sigma)
- 5 M NaCl
- 1 M Tris-HCl (pH 8)
- 10% sodium dodecyl sulfate (SDS)
- Milli-Q water

Method
1. Spot 10 µl of the fixed sample (see *Protocol* 5, step 12) on to the center of a 2.5 × 7.5 cm glass cover slide.

2. Place the slide in a 46°C incubator for 10–15 min to bind the cells to the glass surface.

3. Let the slides air dry in the incubator at 46°C.

4. Prepare the hybridization buffer in accordance with *Table* 3 to achieve the stringency needed for the specific probe.

5. Mix 30 µl of hybridization buffer with 3 µl of FISH probe (50 ng/µl)[a] and transfer on to the area of the dried cells.

6. Place slides horizontally in a 50 ml tube, with the cell side uppermost. Ensure sufficient humidity by placing a piece of napkin (approx. 2 × 5 cm) moistened with 1 ml of hybridization buffer below the slide.

7. Hybridize at 46°C for 1.5 h in the dark.

Notes
[a]The FISH probe used in these experiments is Arch 915 described above.

Table 3. Hybridization solutions for achieving different stringencies under isothermal hybridization conditions

Final amount of formamide (%)	Formamide (µl)	Milli-Q water (µl)	5 M NaCl (µl)	1 M Tris-HCl (µl)	10% SDS (µl)
0	0	1600	360	40	2
5	100	1500	360	40	2
10	200	1400	360	40	2
15	300	1300	360	40	2
20	400	1200	360	40	2
25	500	1100	360	40	2
30	600	1000	360	40	2
35	700	900	360	40	2
40	800	800	360	40	2
45	900	700	360	40	2
50	1000	600	360	40	2
55	1100	500	360	40	2

Protocol 7

Washing after FISH[a]

Equipment and Reagents
- Formamide (Sigma)
- 5 M NaCl
- 1 M Tris-HCl (pH 8)
- 10% SDS
- 0.5 M EDTA (pH 8)
- Milli-Q water
- 50 ml Falcon tube (Becton Dickinson)

Method
1. In a 50 ml Falcon tube, prepare 50 ml of washing buffer at least 30 min before the end of hybridization and pre-heat to 48°C before washing (see *Table 4*).

2. Remove slide(s) from the humid chamber using tweezers.

3. Let a few milliliters of the washing buffer run over the slide to wash off crystals (do not apply directly to the sample, but let the washing buffer run gently over the area with the attached cells).

4. Transfer the slide into the washing chamber (the probed area should be fully submerged in the buffer) and leave for 15 min at 48°C.

5. Briefly submerge slides in 50 ml of Milli-Q water (in a 50 ml Falcon tube) to remove salts present in the washing buffer.

6. Allow the slides to air dry inside the incubator at 46°C.

Notes
[a]Formamide fumes are toxic and we strongly recommend carrying out this work in a fume hood.

Table 4. Washing buffer solutions

Amount of formamide (%) during hybridization	1 M Tris-HCl (μl)	10% SDS (μl)	5 M NaCl (μl)	0.5 M EDTA (μl)
0	1000	50	9000	0
5	1000	50	6300	0
10	1000	50	4500	0
15	1000	50	3180	0
20	1000	50	2150	500
25	1000	50	1490	500
30	1000	50	1020	500
35	1000	50	700	500
40	1000	50	460	500
45	1000	50	300	500
50	1000	50	180	500
55	1000	50	100	500

Solutions were made up to a final volume of 50 ml with Milli-Q water.

Protocol 8

Micromanipulation and amplification of genomic DNA by MDA

Equipment and Reagents
- 22 × 22 mm cover slide (Corning)
- Vaseline
- TE buffer (10 mM Tris-HCl (pH 8); 1 mM EDTA)
- REPLI-g Kit (Qiagen)

Method

1. After the slides have dried (see *Protocol 7*, step 6), carefully apply Vaseline to the border of the cover slide on the side where the probed cells are situated. Mount the cover slide underneath the 3 mm microscope slide so that the cover slide covers the hole of the microscope slide (see *Fig. 4b*). The probed cells on the cover slide should face inwards in the well that is formed.

2. Apply moderate pressure to the cover slide to ensure that it forms a watertight well.

3. Fill the well with TE buffer (pH 8) and mount it in the inverse microscope.

4. Using the microinjector, fill the tip of the microcapillary with sterile TE buffer and manipulate it into the well of the microscope slide.

5. After having located the target cell – as identified by a positive probe signal – manipulate the tip of the microcapillary close to the target cell and suck up the cell in the microcapillary by lowering the pressure inside the microcapillary by means of the microinjector (if necessary apply moderate force to the cell using the microcapillary to loosen the cell from the glass surface).

6. Remove the microcapillary from the micromanipulator and inject the captured cell into 13 µl of sterile TE buffer (pH 8) in a PCR vial.

7. Freeze–thaw the isolated cell five times in TE buffer and spin down the mixture after the final thaw cycle. This step lyses Archaea cells.

8. Carry out steps with the denaturation solution and neutralization solution of *Protocol 2*, which is required for denaturation of the DNA template as well as for cell lysis.

9. Carry out MDA (see *Protocol 3*), following the instructions for a 50 µl MDA reaction in the REPLI-g Kit (Qiagen).

3.5. Bioinformatic aspects of single-cell MDA for genome sequencing and assembly

Bacteria exist in nature as members of complex microbial communities (22). Selection pressure acting on species in these communities may result in the co-evolution of functional specialization and metabolic interdependency (23). Cell signaling and horizontal gene transfer are also noteworthy in these communities, and these information exchanges may cross species boundaries. The study of these phenomena in microbial communities is facilitated by sequencing the genomic DNA of individual member species. For currently unculturable microbes, this goal is unattainable using a conventional genomic DNA sequencing strategy due to an inability to obtain species-specific DNA as a starting material. Unfortunately, more than 99% of microbial species cannot be cultured, or will require extensive research to culture (24). By their nature, thermophiles and other extremophiles will require highly specialized growth media. Among this countless number of unculturable species will be many whose study will contribute greatly to basic research in microbial geochemistry, ecology, and evolution. Many of these unculturable species will also have novel biochemical pathways of great practical importance, e.g. for the production of pharmaceuticals. Whole-genome DNA sequencing and comparative genomics is the most efficient way to identify the genes whose expression products comprise such biochemical pathways. It also yields genome sequence annotation data that can be used to formulate appropriate nutrient media for the laboratory cultivation of important but currently unculturable microbial species (25).

Genomic DNA sequencing strategies and related genome assembly algorithms have been reviewed elsewhere in detail (26) and in shorter monographs (27–30). In the first step of conventional shotgun sequencing, multiple copies of a single genome are obtained from a very large number of cells of a single species. These genomic DNA sequences are fragmented in order to provide sufficiently small pieces of DNA for sequencing. This fragmentation process results in a loss of *linkage information* about sequences in the genome that are *physically linked* (i.e. on the same chromosome or plasmid DNA molecule), but are separated by a distance greater than the fragment size. Collections of these fragments (libraries) are produced, and it is the nature of these libraries that distinguishes different types of conventional shotgun sequencing strategies. Each library has a mean fragment length, and ideally a narrow fragment-length distribution. During library construction, a single fragment is inserted (cloned) into each vector molecule, and only one (vector + insert) recombinant molecule is taken up by a host bacterium during transformation. Each of these clones must be individually propagated *in vivo* in a bacterial host in order to obtain sufficient copies of template DNA for sequencing. Adequate numbers of cloned fragments are then sequenced at their opposite ends, which yields *mate-paired* sequencing reads separated by a known distance. Unless it is required during the finishing step, the central part of a fragment *between* a mate pair is not sequenced. It is easier and cheaper to generate reads using two generic, oppositely oriented primer-binding sites (e.g. that flank a vector cloning site) and rely on read overlap from a high sequencing coverage depth than it is to synthesize custom primers and walk across each insert.

The final phase of all genomic DNA sequencing strategies involves computationally reconstituting the *linkage groups* (chromosome or plasmid DNA molecules) in the genome. Sequence *contigs* (contiguous, gap-free regions) are generated first using overlapping reads. Contigs are then ordered and oriented into *scaffolds* using mate-paired reads that bridge – and define the length of – the unsequenced gap between any given pair of consecutive contigs in a scaffold. Finally, directed sequencing (finishing) is used to fill the known-length gaps within scaffolds and the unknown-length gaps between scaffolds. For bacterial genomes, the reconstituted linkage groups are typically composed of a single chromosome. However, more than one may be present, e.g. the 2.8 Mbp circular and 2.1 Mbp linear chromosomes of *Agrobacterium tumefaciens* (31). In many species (e.g. *A. tumefaciens*), one or more naturally occurring plasmids may also be components of the genome. In *Borrelia burgdorferi* strain B31, a causative agent of Lyme disease, the bacterial genome includes 21 plasmids (32, 33).

MDA has already demonstrated its utility in conventional genomic DNA sequencing strategies (3). It dramatically lowers sequencing costs and waste generation, particularly at the sequencing template preparation step. Single-cell MDA has the potential to make even greater contributions to genomic DNA sequencing strategies and other types of genome analysis. This potential should be enhanced by recent advances in three complementary research fields: single-cell isolation; microfluidic device design and fabrication (especially those developed for use with single cells); and new, nonelectrophoretic DNA sequencing technologies (34). For example, although single cells are typically not lysed quantitatively using scaled-down versions of traditional lysis protocols, microfluidic devices have been developed that can significantly increase the efficiency of this procedure. The use of such devices should help minimize loci representation anomalies during single-cell MDA reactions.

Single-cell MDA of unculturable microbes would be a very useful adjunct to metagenomic shotgun sequencing strategies. The metagenomic approach seeks to accomplish the parallel assembly of multiple genomes of unculturable microbes using DNA extracted from field samples of microbial communities (22, 35). Some metagenomic shotgun sequencing strategies include steps for the partial isolation of some types of genomes, e.g. via the size separation of bacterial cells from viruses by filtration. However, in most cases the DNA extracted from the field sample or one of its fractions has a *complexity* (total length of nonrepetitive DNA from all genomes) that is unknown, difficult to obtain in a reproducible manner, and potentially very large. As a result, only a modest number of complete genome assemblies may be obtained relative to the total sequencing effort (9). Genome assembly problems may arise because the field sample contained too many different species, resulting in inadequate sequence coverage per genome (36). The field sample could also have contained too many related species or strains, resulting in an inability to resolve multiple genome assemblies because of closely related sequences. The degree of sequence relatedness that distinguishes species, or distinguishes related strains of the same species, is often unclear; and even isolates of the same strain may display significant genomic diversity (37). The DNA complexity of a metagenomic field sample may also be dominated by large numbers of small genomes such as from temperate bacteriophage (38). These and other small invasive genetic elements (e.g. transposons) may not be of phenotypic interest, but if they are integrated into the genomes of multiple species or strains in the field sample their presence will complicate genome assembly. Thus, the feasibility of the metagenomic strategy appears to be inversely related to the DNA sequence complexity of the field sample (39). Finally, even if they can be assembled, there is no obvious solution offered by the metagenomic strategy to the assignment of nonchromosomal linkage groups such as plasmids to individual microbial genomes.

Single-cell MDA of unculturable microbes yields genomic DNA sequence data that could complement metagenomic sequencing strategies in several ways. There is a one-to-one relationship between a genome and a cell, and thus several of the problems encountered during the genome assembly phase of a metagenomic project could be ameliorated by the concomitant use of sequence data obtained using single-cell MDA on individual cells present in the field sample. Even partial genomic sequence obtained from single-cell MDA could enable or validate the assignment of genes to species-specific linkage groups and the assignment of multiple linkage groups (i.e. plasmids and chromosomes) to single species. It could also help define or confirm putative new species and allow the design of species-specific probes for PCR, FISH, and microarray analyses. Sequence data obtained using single-cell MDA could also help relate sequence information to phenotypic characteristics observed by microscopy and enable more sophisticated phenotypic fractionation of the microbes in the field sample prior to sequencing. The detection of novel phenotypes could itself be at the single-cell level, using FISH labeling and micromanipulation techniques (see *Protocols 5–8*). This would be particularly useful for the isolation of genomes that may be minor contributors to the DNA complexity of a microbial community, but which produce gene products that are of major phenotypic importance.

There are other areas of prokaryotic and eukaryotic genome research where a single-cell MDA approach may prove useful. For example, single-cell MDA may be used in *disjoint* genomic DNA sequence analyses. A single-cell lysate could be *divided* into several equal-sized aliquots, with each containing fragments representing a unique subset of the genome, and these aliquots of template DNA could be used in *separate* MDA reactions. The complexity of the genome would be disjointly partitioned among these aliquots. If a large number of aliquots were used, individual copies of multi-copy (e.g. 16S rRNA) bacterial genes would be more likely to end up in different disjoint aliquots. Under appropriate conditions, such as when cells are not actively dividing, it is reasonable to assume that a single copy of a genome is present in a cell used for disjoint MDA reactions. Genomic DNA sequence data obtained using such single-cell disjoint MDA reactions could help resolve genome sequence assembly errors or ambiguities that may have arisen due to multi-copy genes, closely related paralogs, and other types of repeated sequences. The disjoint MDA approach could also enable the sequencing and assembly of haplotypes when a single diploid eukaryotic cell is used. The length of each assembled haplotype would have a theoretical upper bound equal to the *pre-amplification* length of the relevant DNA fragment in the single-cell lysate (i.e. *before* the disjoint MDA reaction). Subsequent library construction (e.g. of a mini-library *specific* for a single-cell disjoint MDA reaction product) would not prevent the assembly of most haplotypes. Thus, in the diploid case, if n is the number of equal-sized disjoint aliquots used, a single (haplotype-bearing) fragment f_m would have a probability of $1/n$ of ending up in the same disjoint aliquot as the corresponding homologous fragment f_p and hence the fraction of sequence data that could be assembled as haplotypes would be $(n-1)/n$.

Single-cell disjoint MDA reaction products could also be used in hybridization-based analytical procedures, e.g. after spotting on an array. If an oligonucleotide probe for a known single-copy gene hybridized to multiple spots (i.e. multiple disjoint aliquots), it may, depending on the ploidy of the single cell used, indicate the presence of multiple copies of the gene in the genome. Similarly, single-copy oligonucleotide probes specific for the ends of scaffolds may be used to identify specific single-cell disjoint MDA reaction products that could be used for further sequencing to extend and possibly close gaps between scaffolds, i.e. during the finishing step of genome assembly. If a diploid eukaryotic cell is used, probes specific for alleles whose interloci distance is less than the length of the relevant DNA fragment in the single-cell lysate may yield information about haplotypes without having to resort to DNA sequence analysis.

A single-cell disjoint MDA approach may be particularly well suited for use with new, nonelectrophoretic DNA sequencing technologies. Single-cell disjoint MDA products may be an appropriate substrate for sequencing by hybridization on oligonucleotide microarrays (40) or other novel sequencing technologies. Many of these new technologies produce short-read DNA sequence data. These reads have lengths of approximately 100–200 nt and typically are not produced as mate pairs. The reduced complexity of the DNA present in single-cell disjoint MDA reaction products would be highly desirable for the assembly of such short-read data, as the absence of mate-pair information currently limits the short-read approach to assemblies of small size (41). Finally, it should be noted that, using current single-cell lysis protocols, bacterial cells are typically not lysed quantitatively. There is dropout of a certain percentage of the genome and loci representation anomalies (see *Table 2*), but it is random for each individual single-cell lysate (see *Fig. 2*). A compensatory tactic would be to select additional single cells of the same species (based on morphological criteria for some species, or in other cases by FISH using species-specific rRNA probes) for a set of single-cell disjoint MDA reactions.

The extraordinary sensitivity of MDA requires that rigorous steps be taken to prevent the introduction of contaminating DNA, e.g. from reagents, aerosols, or other sources. Some chimeric MDA products may also be observed (Andreas Schirmer and Daniel Santi, Kosan Biosciences, California, USA, personal communication). The deleterious effect of contaminating DNA sequences and chimeric MDA product generation on assembly quality can be minimized by the appropriate use of functionality found in most genome assemblers, e.g. the screening of known contaminating sequences (as is typically used for vector sequences). As more experience is gained with the use of single-cell MDA products for genomic DNA sequencing, machine learning techniques may also be able to use laboratory-specific historical data on contaminating sequences to refine the screening of these sequences further. The sequencing of single-cell MDA-generated products to additional coverage depth is also an option to minimize the effects of these anomalies. Finally, contaminating DNA sequences and chimeric fragment generation are generic problems associated with any genome sequencing strategy. Even if they escape removal at earlier stages (e.g. by screening), these anomalies will typically end up as isolated contigs that are eventually rejected from the assembly due to their inability to be incorporated into scaffolds.

Comparative genomic studies of intraspecies variation, or studies of interspecies and intraspecies exchanges of genetic information, are very important areas of microbiological research. The validity of these studies depends on the accuracy and completeness of the genomic DNA sequence, and the extent and accuracy of its annotation. Bacteria often display significant intraspecies variation (37) and there is strong theoretical reasoning and empirical evidence that the distribution of intraspecies variation among individuals is uneven (42). Thus, obtaining sequence data from several individual members of a species that has significant variation facilitates the annotation of the genome of that species, as regions of low sequence variation are often functionally constrained (43). The microscopic and micromanipulation procedures described here also lend themselves to the isolation of individual microbial cells of a given species that exhibit morphological or other observable or measurable forms of phenotypic variation. These single-cell analytical and isolation capabilities, combined with the use of a single-cell MDA genomic sequencing strategy, open up exciting new opportunities in microbial genetics.

4. REFERENCES

1. Dean FB, Nelson JR, Giesler TL & Lasken RS (2001) *Genome Res.* **11**, 1095–1099.
★★★ 2. Dean FB, Hosono S, Fang L, *et al.* (2002) *Proc. Natl. Acad. Sci. U. S. A.* **99**, 5261–5266. – *The first report of WGA by MDA including a detailed analysis of reaction characteristics.*
★★★ 3. Detter JC, Jett JM, Lucas SM, *et al.* (2002) *Genomics*, **80**, 691–698. – *The first publication of library construction and DNA sequencing from MDA reactions.*
4. Nelson JR, Cai YC, Giesler TL, *et al.* (2002) *Biotechniques* **Suppl.**, 44–47.
5. Lage JM, Leamon JH, Pejovic T, *et al.* (2003) *Genome Res.* **13**, 294–307.
★★★ 6. Raghunathan A, Ferguson HR, Bornarth CJ, Driscoll M & Lasken RS (2005) *Appl. Environ. Microbiol.* **71**, 3342–3347. – *The first report of MDA performed using single bacterial cells.*
★★★ 7. Ishøy T, Kvist T, Westermann P & Ahring BK (2005). *Appl. Microbiol. Biotechnol.* **In press.** – *Excellent detailed description and validation of the micromanipulation methods in this chapter.*
8. Schloss PD & Handelsman J (2003) *Curr. Opin. Biotechnol.* **14**, 303–310.
9. Venter JC, Remington K, Heidelberg JF, *et al.* (2004) *Science*, **304**, 66–74.
10. Hosono S, Faruqi AF, Dean FB, *et al.* (2003) *Genome Res.* **13**, 954–964.
11. Jeyaprakash A & Hoy MA (2004) *J. Invertebr. Pathol.* **86**, 111–116.
12. Gadkar V & Rillig MC (2005) *FEMS Microbiol. Lett.* **242**, 65–71.
13. Schmeisser C, Stockigt C, Raasch C, *et al.* (2003) *Appl. Environ. Microbiol.* **69**, 7298–7309.
14. Bianchi A & Giuliano L (1996) *Appl. Environ. Microbiol.* **62**, 174–177.
15. Browning I & Lockingen LS (1952) *Science*, **115**, 646–647.
16. Manhoff LJ & Johnson MW (1950) *Science*, **112**, 76–77.
17. Nygaard G (1949) *Science*, **110**, 165–166.
18. Huber R (1999) *Biospektrum*, **5**, 289–291.
★★★ 19. Fröhlich J & König H (1999) *Syst. Appl. Microbiol.* **22**, 249–257. – *Description of an alternative micromanipulation setup.*
20. Burmolle M, Hansen LH, Oregaard G & Sorensen SJ (2003) *Microb. Ecol.* **45**, 226–236.
21. Stahl DA & Amann R (1991) In: *Nucleic Acid Techniques in Bacterial Systematics*, pp. 205–248. Edited by E Stackebrandt & M Goodfellow. John Wiley & Sons, UK.
22. Riesenfeld CS, Schloss PD & Handelsman J (2004) *Annu. Rev. Genet.* **38**, 525–552.
23. Shapiro JA (1998) *Annu. Rev. Microbiol.* **52**, 81–104.
24. Rappe MS & Giovannoni SJ (2003) *Annu. Rev. Microbiol.* **57**, 369–394.

25. Bhattacharyya A, Stilwagen S, Reznik G, *et al.* (2002) *Genome Res.* **12**, 1556–1563.

★★★ 26. Pop M (2004) In: *Advances in Computers*, Vol. 60, pp. 193–248. Edited by M Zelkowitz. Elsevier, Amsterdam. – *Review of shotgun sequencing and assembly.*

27. Gibbs RA & Weinstock GM (2003) *Cold Spring Harb. Symp. Quant. Biol.* **68**, 189–194.

28. Green ED (2001) *Nat. Rev. Genet.* **2**, 573–583.

★★★ 29. Myers G (1999) *Comput. Sci. Eng.* **1**, 33–43. – *Introduction to shotgun sequencing and assembly.*

30. Frangeul L, Nelson KE, Buchrieser C, Danchin A, Glaser P & Kunst F (1999) *Microbiology,* **145**, 2625–2634.

31. Wood DW, Setubal JC, Kaul R, *et al.* (2001) *Science,* **294**, 2317–2323.

32. Fraser CM, Casjens S, Huang WM, *et al.* (1997) *Nature,* **390**, 580–586.

33. Rosa PA, Tilly K & Stewart PE (2005) *Nat. Rev. Microbiol.* **3**, 129–143.

34. Shendure J, Mitra RD, Varma C & Church GM (2004) *Nat. Rev. Genet.* **5**, 335–344.

35. Streit WR & Schmitz RA (2004) *Curr. Opin. Microbiol.* **7**, 492–498.

36. Tringe SG, von Mering C, Kobayashi A, *et al.* (2005) *Science,* **308**, 554–557.

37. Cohan FM (2002) *Annu. Rev. Microbiol.* **56**, 457–487.

38. Breitbart M, Felts B, Kelley S, *et al.* (2004) *Proc. Biol. Sci.* **271**, 565–574.

39. Tyson GW, Chapman J, Hugenholtz P, *et al.* (2004) *Nature,* **428**, 37–43.

40. Drmanac R, Drmanac S, Chui G, *et al.* (2002) *Adv. Biochem. Eng. Biotechnol.* **77**, 75–101.

41. Chaisson M, Pevzner P & Tang H (2004) *Bioinformatics,* **20**, 2067–2074.

42. Rauch EM & Bar-Yam Y (2004) *Nature,* **431**, 449–452.

43. Boffelli D, Weer CV, Weng L, *et al.* (2004) *Genome Res.* **14**, 2406–2411.

CHAPTER 10

Genome amplification tolerant to sample degradation: application to formalin-fixed, paraffin-embedded specimens

G. Mike Makrigiorgos

Dana-Farber/Brigham and Women's Cancer Center, Level L2, Radiation Therapy, 75 Francis Street, Boston, Massachusetts 02115, USA

1. INTRODUCTION

New technologies for analyzing gene expression and genomic DNA differences are expected to lead to profound changes in the detection, prognosis and therapy of cancer in the near future. For example, increasingly tight associations between genomic/gene-expression signatures and the biology and outcome of prostate cancer (1–5) and breast cancer (6–12) have been revealed by DNA microarray technology. However, prospective clinical studies to validate the new molecular knowledge can be very expensive and usually take several years to yield their conclusions due to the required follow-up. As an alternative, retrospective studies on existing specimens are a very useful option. Formalin-fixed, paraffin-embedded (FFPE) tissue biopsies present in the archives of departments of pathology represent an extensive source of morphologically defined specimens donated by patients for whom complete clinical data are already available. Coupled to the recent technological developments in genomics, these FFPE archives provide a unique resource for retrospective analysis to accelerate the quest for clinically relevant biomarkers, which can then guide the design of major, prospective studies in a more efficient manner. Nevertheless, several technical hurdles persist. Firstly, DNA and RNA in prostate FFPE tissue biopsies are frequently moderately to highly degraded (13). Secondly, since cancer specimens are often heterogeneous, containing stromal cells, infiltrating leukocytes, and tumor cells that harbor a range of genetic profiles (14), meaningful analysis must often be

Whole Genome Amplification: *Methods Express* (S. Hughes and R. Lasken, eds.)
© Scion Publishing Limited, 2005

coupled to microdissection in order to remove homogeneous cell populations (15). Thirdly, analysis of genetic changes in FFPE tumors using microarrays generally requires microgram quantities of pure tumor DNA (16, 17). To address these problems many laboratories couple laser-capture microdissection (LCM), which allows removal of minute amounts of homogeneous tissue from tumors (15, 18, 19), with WGA of the extracted nucleic acids in order to generate sufficient DNA for microarray screening. However, LCM itself can produce additional nucleic acid damage that compounds the existing DNA/RNA damage in FFPE specimens (19). LCM of heterogeneous FFPE specimens, as well as fresh specimens, would significantly benefit from being coupled to a WGA method tolerant to sample degradation.

A new adaptation of isothermal rolling-circle amplification (RCA), termed restriction and circularization-aided RCA (RCA–RCA) (20), can overcome the problems associated with significantly degraded FFPE samples. In addition, RCA–RCA retains the distinction among corresponding genes when a 'target' and a 'control' genome/transcriptome are amplified.

2. METHODS AND APPROACHES

2.1. Principles of RCA–RCA

Formalin fixation of tissue results in DNA strand breaks, base damage, and DNA–protein crosslinks, all of which inhibit amplification (13, 21). The principle of RCA–RCA is that fragmenting the genome with an appropriate restriction enzyme, which cuts at least twice within a pre-existing DNA fragment, generates DNA that can self-ligate/circularize (see *Fig. 1*). Following circularization and degradation of noncircular DNA using exonuclease, the circles are denatured to enable initiation of exponential, hyper-branched multiple displacement amplification (MDA) (22) using random hexamer primers and φ29 DNA polymerase.

Depending on the degree of FFPE sample degradation, a restriction enzyme that cuts with lower or higher frequency can be chosen without further modification of the procedure. The choice of enzyme guarantees that the highest possible fraction of the genome is captured in circles, including small fragments. Although self-circularization of DNA fragments <250 bp in size is inefficient (23), we found (20, 24) that cross-ligation of the smaller DNA fragments forms larger fragments that subsequently circularize. We recently demonstrated the utility of RCA–RCA for the unbiased amplification of degraded formalin-fixed DNA samples on a genome-wide level, as well as for amplification of cDNA (20).

2.2. Advantages of RCA–RCA

Multiple protocols for PCR-based WGA have been described in the literature, including primer-extension pre-amplification (25), degenerate-oligonucleotide-primed PCR (26), Alu PCR (27), and ligation-mediated amplification (17).

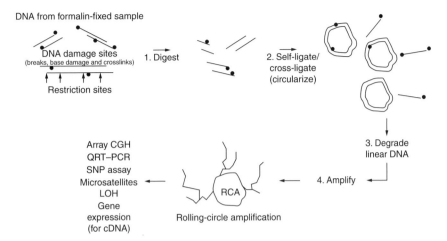

Figure 1. Outline of WGA of partially degraded FFPE samples via RCA–RCA.
DNA damage sites (●) are indicated. The same approach may also be used to amplify
intact or degraded genomic DNA or cDNA. The amplified products are good substrates
for quantitative real-time (QRT) PCR, array CGH, loss of heterozygosity (LOH) studies,
and single nucleotide polymorphism (SNP) and microsatellite analysis.

Depending on the technique, all are capable of amplifying DNA extracted from
as little as a single cell up to a few hundred cells. However, the introduction of
PCR bias during amplification is a concern, as the quantitative relationship
between alleles before and after amplification can be affected. In an attempt to
overcome PCR-introduced bias during whole genome/transcriptome
amplification, we recently developed balanced PCR (28). Balanced PCR was
shown to allow unbiased screening for gene expression (28) and genomic (29)
applications. However, since PCR amplifies only a minor portion of the entire
genome (16), vital genetic regions may be lost. Lizardi and co-workers
introduced RCA (30), an approach that subsequently led to an isothermal WGA
method known as MDA (31, 32). MDA operates on long DNA templates and
produces amplified genomic DNA starting from intact genomes obtained from
cells or fresh tissue, and is currently considered the method of choice for several
applications (25, 33–35). However, the amplification efficiency of MDA
diminishes as the size of the DNA strands decreases, which is a major problem
for amplification of FFPE DNA (32). RCA–RCA combines the advantages of PCR
and MDA as it works with fragmented DNA while providing an almost complete
genome coverage (20).

In vitro transcription T7-based methods have been successfully adapted to the
quantitative amplification of RNA (36–39) prior to microarray screening. In
contrast, RCA–RCA amplifies cDNA molecules instead of RNA and utilizes
processes that are robust and faster than the T7 method, while also being of lower
cost. The single format used both for expression (cDNA) and for genomic profiling
(DNA) from FFPE samples makes RCA–RCA more generally applicable than the T7
method.

3. RECOMMENDED PROTOCOLS

Protocol 1

DNA circularization

The DNA used in these experiments was extracted either from cultured cell lines or from FFPE specimens[a].

Equipment and reagents

- *Nla*III restriction enzyme (10 units/µl) (New England Biolabs)
- T4 DNA ligase buffer (New England Biolabs)
- λ Exonuclease (5 units/µl) and the accompanying λ exonuclease buffer (New England Biolabs)
- Exonuclease I (20 units/µl) (New England Biolabs)
- QIAquick PCR Purification Kit (Qiagen)

Method

1. Digest the genomic DNA (20–50 ng) or cDNA with 0.5 µl *Nla*III (10 units/µl) at 37°C for 2 h in 10 µl of T4 DNA ligase buffer.

2. Heat the sample at 65°C for 20 min to inactivate *Nla*III.

3. Add 0.5 µl T4 DNA ligase and circularize the fragmented DNA in a volume of 15 µl, in T4 ligase buffer, at room temperature for 2 h.

4. Inactivate the ligase by incubating at 65°C for 10 min.

5. Eliminate linear DNA by adding 1.2 µl of λ exonuclease (5 units/µl) and 0.3 µl exonuclease I (20 units/µl) in a volume of 25 µl. Make up the final volume to 25 µl with exonuclease buffer.

6. Incubate at 37°C for 1 h.

7. Purify the circularized DNA using the QIAquick PCR Purification Kit and elute in 35 µl of water[b].

Notes

[a]To extract DNA from FFPE specimens, established (21) approaches involving rinsing of paraffin sections via xylene/centrifugation/ethanol, washing out of the fixative with phosphate-buffered saline and extraction of DNA/total RNA using the QIAamp DNA Kit (Qiagen) can be used.
[b]Following DNA circularization, exonuclease digestion, and purification, approximately 30–50% of the original DNA is recovered.

Protocol 2

Amplification of circularized DNA

Equipment and reagents
- Random hexamers (400 ng/μl) (Sigma)
- φ29 DNA polymerase (10 units/μl) and accompanying 10× φ29 DNA polymerase buffer (New England Biolabs)
- 100× Bovine serum albumin (BSA; 10 mg/ml) (Sigma)
- Binding buffer (400 mM Tris-HCl (pH 8.0); 160 mM KCl; made up in RNase/DNase-free, millipore-purified water)
- 2.5 mM dNTPs (Applied Biosystems)
- 20% Dimethyl sulfoxide (DMSO) (Sigma)
- T4 gene 32 protein (20 ng/μl) (Amersham Biosciences)
- Reference human male genomic DNA (G147A) (Promega)
- NlaIII restriction enzyme (10 units/μl) (New England Biolabs)

Method

1. Mix 4 μl of circular DNA with 0.5 μl of random hexamers[a] and 0.5 μl of binding buffer and denature at 95°C for 4 min[b].

2. Following denaturation, amplify the DNA using 0.3 μl of φ29 DNA polymerase, complemented with 2 μl of 10× φ29 DNA polymerase buffer, 0.2 μl of 100× BSA, 3.2 μl of 2.5 mM dNTPs, 1 μl of 20% DMSO, and 1 μl of 20 ng/μl T4 gene 32 protein in a final volume of 20 μl at 30°C for 6–16 h[c].

3. Inactivate φ29 DNA polymerase at 65°C for 5 min[d,e,f,g].

4. Digest the viscous RCA–RCA product with 2.5 μl of NlaIII (10 units/μl) at 37°C for 3 h in a volume of 100 μl (make up to 100 μl with nuclease-free water) to reduce the DNA size prior to performing downstream assays such as real-time PCR from genomic DNA or cDNA, array comparative genomic hybridization (CGH), or mutation/microsatellite detection (20).

5. Alternatively, instead of step 4 (NlaIII digestion), heat the amplified sample at 95°C for 7–10 min to reduce the DNA size.

Notes

[a]In (22), the random hexamers used were modified to resist 3′→5′ exonuclease degradation during MDA. Commercial kits for MDA (GenomiPhi, Amersham; REPLI-g, Qiagen) also use modified hexamers. Modified hexamers can be obtained from Integrated DNA Technologies. However, in our laboratory, unmodified hexamers from Sigma were used for RCA–RCA (20). We found that the unmodified hexamers performed adequately for RCA–RCA using the recommended buffer conditions.

[b]The denaturation step can be omitted, as φ29 DNA polymerase has the ability to start displacement using a double-stranded template.

[c]An overnight 16 h reaction is often convenient.

[d]A negative control for the reaction (i.e. identical sample without template genomic DNA added) may also be included and run in parallel.

[e]Unlike MDA, there are no nonspecific amplification products using RCA–RCA, possibly due to the buffer composition used (20). Therefore, the negative control should yield no detectable 'smear' or 'bands' on agarose gel electrophoresis (see *Fig. 2b*).

[f]In view of the absence of nonspecifically amplified products, the success of RCA–RCA can be verified by a simple fluorometric DNA quantitation using PicoGreen (20). Alternatively, Taqman real-time PCR of a housekeeping gene such as glyceraldehyde phosphate dehydrogenase (GAPDH) can be used to evaluate amplification.

[g]Overall amplification of 1000–8000-fold is usually obtained with RCA–RCA, depending on DNA quality. Highly degraded FFPE samples generally yield lower amplifications than intact samples.

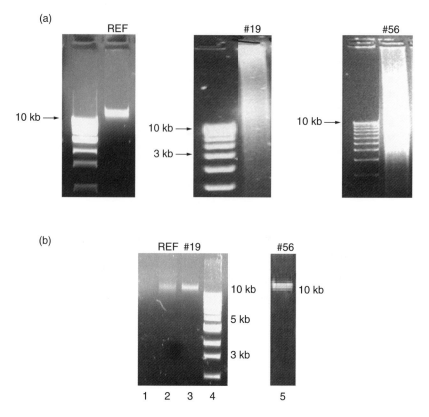

Figure 2. Agarose gel electrophoresis of RCA–RCA-amplified, intact, and FFPE genomic DNA.
(*a*) Agarose gel electrophoresis of reference DNA (REF; Promega) and DNA from two FFPE samples, #19 and #56. The left lane contains DNA size markers. (*b*) RCA–RCA products from 50 ng of intact reference DNA (*lane 2*), 50 ng of #19 DNA (*lane 3*) and 50 ng of #56 DNA (*lane 5*). *Lane 1* was a negative control (using water instead of template DNA) and *lane 4* contained DNA size markers.

3.1. Examples of results

3.1.1. Real-time PCR examination of RCA–RCA-amplified, intact, and FFPE genomic DNA

DNA extracted from formalin-fixed (40), paraffin-embedded tissue was used for amplification via RCA–RCA. The electropherograms in *Fig. 2*(*a*) depict an increasing degree of DNA degradation, relative to reference DNA, in two FFPE samples of low and modest degradation (#19 and #56, respectively). Following RCA–RCA of the reference and FFPE samples, high-molecular-weight DNA was obtained from all three samples (see *Fig. 2b*).

In order to examine the ability of RCA–RCA to generate unbiased WGA, we utilized DNA from BT474 cells (ATCC HTB-20), which are known to contain well-

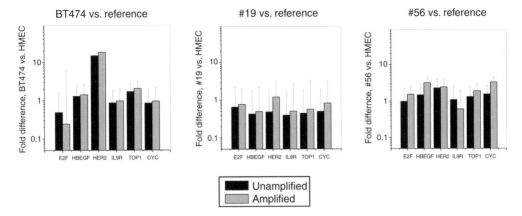

Figure 3. Quantitative PCR (Taqman assay)-based comparison of gene copy number ratios before and after RCA–RCA.
DNA was extracted from BT474 cells, FFPE samples #19 and #56 and compared with reference DNA (Promega). The relative (ΔΔCt) threshold method was used to compare relative gene copy number before amplification and after RCA–RCA using the GAPDH housekeeping gene for normalization. The results are presented as copy number ratio relative to co-amplified reference DNA. The genes tested were transcription factor E2F (E2F), heparin-binding epidermal growth factor-like growth factor precursor (HBEGF), human epidermal growth factor receptor 2 (HER2), interleukin-9 receptor (IL9R), topoisomerase 1 (TOP1), cytochrome c (CYC). HMEC, human mammary epithelial cells.

characterized genetic amplifications and deletions (29, 41). *Fig. 3* demonstrates that the BT474 cell DNA amplified by RCA–RCA retains the previously described (41) HER2 amplification when amplified and unamplified samples are compared (Pearson correlation coefficient, R^2=0.98). The relative ratios for the same genes in DNA from FFPE samples #19 and #56 relative to the reference DNA are also in concordance (R^2=0.56–0.65) when amplified and unamplified samples are compared. The error bars in these figures represent real-time PCR performed from three independent RCA–RCA experiments. The data indicate that RCA–RCA can amplify FFPE samples of modest or significant degradation and retain the copy number variation. However, we have observed that, as degradation of the examined FFPE samples increases further, the amplification obtained for selected genes is reduced and the copy number variation becomes more difficult to retain (20). It may be possible to recover information from highly degraded FFPE samples using a more frequently cutting restriction enzyme than *NlaIII* in the RCA–RCA protocol.

3.1.2. Array CGH screening of RCA–RCA-amplified genomic DNA

To validate the application of RCA–RCA-amplified DNA for array CGH studies, we used 20 ng of DNA extracted from BT474 cells for amplification. We then screened the product against reference genomic DNA that had been similarly amplified. The RCA–RCA products were screened in two duplicate independent experiments

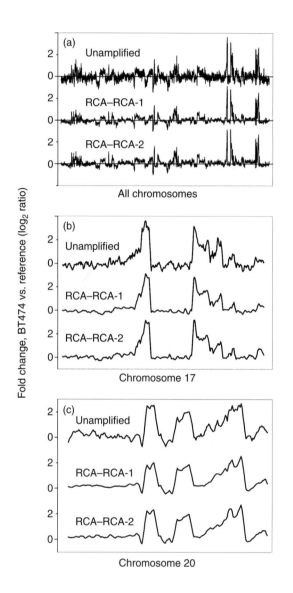

Figure 4. Array CGH screening for RCA–RCA-amplified BT474 genomic DNA vs. reference DNA.

'Unamplified' represents the array CGH results from unamplified BT474 genomic DNA vs. reference DNA (Promega). RCA–RCA-1 and RCA–RCA-2 are duplicate RCA–RCAs performed in two separate experiments. (*a*) Comparison for all 23 chromosomes. (*b*) Comparison for chromosome 17. (*c*) Comparison for chromosome 20. Five-nearest-neighbor smoothing was used for these data, performed using the ORIGIN 7.5 software suite (OriginLab). Nearest-neighbor smoothing takes the average of a certain number (five points in this case) of data points around each point in your data and replaces that point with the new average value. The $\log_2$ ratio corresponds to fold differences in gene dosage, whereby a $\log_2$ ratio of 1 is equal to a twofold difference; a $\log_2$ ratio of 2 is equal to a fourfold difference, etc.

using the Agilent Human 1 cDNA microarrays. Unamplified BT474 DNA (4 µg) was also screened directly on microarrays vs. unamplified reference DNA. The genomic profiles from the duplicate RCA–RCA products showed a pattern very similar to that from unamplified samples (see *Fig. 4a*). BT474 cancer cells contain well-known multiple amplification regions in chromosomes 17 and 20, which are depicted in more detail in *Fig. 4*(*b* and *c*). For both chromosomes, regions of

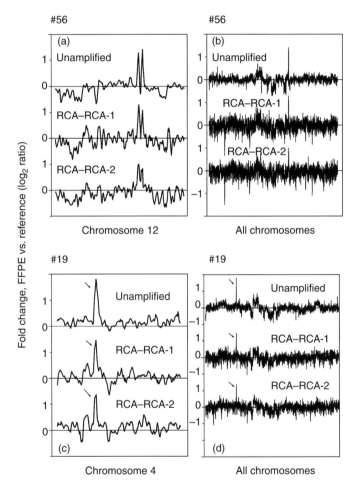

Figure 5. Array CGH screening for RCA–RCA-amplified genomic DNA from FFPE samples vs. reference DNA.

'Unamplified' represents the array CGH results from FFPE samples #19 or #56 vs. reference DNA (Promega). RCA–RCA-1 and RCA–RCA-2 are duplicate RCA–RCAs performed in two separate experiments. (*a, c*) Comparison for chromosome 12 for FFPE sample #56 (*a*) and for chromosome 4 for FFPE sample #19 (*c*). (*b, d*) Comparison for all 23 chromosomes. Five-nearest-neighbor smoothing was used for these data, performed using the ORIGIN 7.5 software suite (OriginLab). See *Fig. 4* legend for explanation of nearest-neighbor smoothing.

genetic amplification observed in unamplified BT474 DNA were also observed in the duplicate amplified samples. Data analysis demonstrated that the concordance of gene-dosage profiling between amplified and unamplified DNA increases if nearest-neighbor averaging is employed for data smoothing. We found that increased data smoothing improves R^2 but also results in a loss of genome-wide resolution. For example, for chromosome 17, R^2 improves from 0.86 to 0.95 when averaging is applied by two or 12 nearest neighbors, respectively. Concomitantly, the array CGH resolution decreases from 100 kbp to 1.2 Mbp average spacing along a chromosome.

Array CGH experiments were also performed for the FFPE samples #19 and #56. Array CGH data obtained for unamplified #19 and #56 samples vs. reference genomic DNA (4 µg) were compared with those obtained for RCA–RCA-amplified #19 and #56 samples vs. amplified reference genomic DNA (20 ng DNA used for amplification). RCA–RCA of DNA from samples #19 and #56 generated array CGH profiles similar to those obtained when unamplified FFPE samples were directly screened on cDNA microarrays. *Fig. 5(a–d)* demonstrates that RCA–RCA, performed in duplicate independent experiments, reproduced the main amplification (more than twofold) regions in samples #19 and #56.

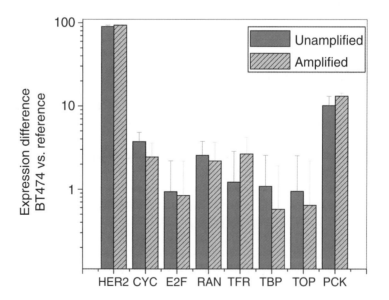

Figure 6. Detection of relative gene expression of BT474 cells vs. reference human mammary epithelial cells before and after RCA–RCA of cDNA, using Taqman real-time PCR.

cDNA used for RCA–RCA was obtained from reverse transcription of ~25 ng of total RNA starting material. The genes tested were human epidermal growth factor receptor 2 (HER2), cytochrome c (CYC), transcription factor E2F (E2F), Ras-related nuclear protein (RAN), transferrin receptor (TFR), TATA box-binding protein (TBP), topoisomerase 1 (TOP), and phosphoenolpyruvate carboxykinase (PCK).

3.1.3. Application of RCA–RCA to cDNA amplification

To examine whether RCA–RCA can be used for the unbiased amplification of cDNA, we generated double-stranded cDNA from 25 ng of total RNA obtained from reference mammary epithelial cells and BT474 cells. The cDNA was used either directly in a Taqman quantitative (Q–PCR) to examine gene dosage for eight genes, or was first amplified via RCA–RCA and then examined via Q–PCR using the same genes. *Fig. 6* (average of three independent RCA–RCA experiments) demonstrates the upregulation in the *HER2* and *PCK1* genes as well as a good retention (R^2=0.99) of the relative gene expression between reference and BT474 cells following RCA–RCA of total cDNA. It should be noted that, with the present approach, end fragments of cDNA sequences would be lost, since only sequences between two restriction sites will be amplified. Nevertheless, the data indicate that a single RCA–RCA protocol can be used for the unbiased amplification of intact or fragmented genomic DNA as well as cDNA.

In conclusion, RCA–RCA is an attempt to overcome the problem of DNA fragmentation in archived tissue specimens by performing self-ligation prior to MDA. The technique performs well for FFPE samples with modest to significant degradation. In addition, following RCA–RCA, a range of assays (PCR-based assays for genotyping or gene dosage, and microarrays) can be reliably applied. It will be interesting to determine whether, by modifying the protocol, it will also be possible to recover information from highly degraded FFPE samples.

4. TROUBLESHOOTING

It is sometimes desirable to use the RCA–RCA product for downstream PCR amplification of regions that contain one or more *Nla*III sites. Because the first RCA–RCA step entails digestion with *Nla*III, PCR amplification from the RCA–RCA product is not possible. In this case, an alternative restriction enzyme should be selected for RCA–RCA (e.g. *Taq*I or *Dpn*II), such that there are no restriction sites in the PCR region(s) of interest. Because the RCA–RCA protocol utilizes a single-tube approach for digestion, ligation, and exonuclease treatment (in ligation buffer), it is important first to verify that the enzyme chosen performs efficiently in ligation buffer. In addition, it is recommended that when enzymes other than *Nla*III are used, RCA–RCA is followed by heating at 95°C, instead of re-digestion, to reduce the viscosity of the sample prior to downstream applications.

Acknowledgements

The assistance of my collaborators Gang Wang, Elizabeth Maher, and Lynda Chin is gratefully acknowledged. Funding for this work was provided in part by the JCRT Foundation, by DOD grant BC020504, by NIH grant 1 R21 CA111994-01, NIH P50 CA 93683, and by the Arthur and Rochelle Belfer Cancer Genomics Center.

5. REFERENCES

1. Lapointe J, Li C, Higgins JP, *et al.* (2004) *Proc. Natl. Acad. Sci. U. S. A.* **101**, 811–816.
2. Nelson PS (2002) *Ann. N. Y. Acad. Sci.* **975**, 232–246.
3. Best CJ, Leiva IM, Chuaqui RF, *et al.* (2003) *Diagn. Mol. Pathol.* **12**, 63–70.
4. Elek J, Park KH & Narayanan R (2000) *In Vivo*, **14**, 173–182.
5. Singh D, Febbo PG, Ross K, *et al.* (2002) *Cancer Cell*, **1**, 203–209.
6. van 't Veer LJ, Dai H, van de Vijver MJ, *et al.* (2002) *Nature*, **415**, 530–536.
7. Perou CM, Jeffrey SS, van de Rijn M, *et al.* (1999) *Proc. Natl. Acad. Sci. U. S. A.* **96**, 9212–9217.
8. Perou CM, Sorlie T, Eisen MB, *et al.* (2000) *Nature*, **406**, 747–752.
9. Sorlie T, Perou CM, Tibshirani R, *et al.* (2001) *Proc. Natl. Acad. Sci. U. S. A.* **98**, 10869–10874.
10. Alizadeh AA, Ross DT, Perou CM & van de Rijn M (2001) *J. Pathol.* **195**, 41–52.
11. Ross DT & Perou CM (2001) *Dis. Markers*, **17**, 99–109.
12. West M, Blanchette C, Dressman H, *et al.* (2001) *Proc. Natl. Acad. Sci. U. S. A.* **98**, 11462–11467.
13. Lewis F, Maughan NJ, Smith V, Hillan K & Quirke P (2001) *J. Pathol.* **195**, 66–71.
14. Shah RB, Mehra R, Chinnaiyan AM, *et al.* (2004) *Cancer Res.* **64**, 9209–9216.
15. Rubin MA (2002) *Science*, **296**, 1329–1330.
16. Lucito R, Nakimura M, West JA, *et al.* (1998) *Proc. Natl. Acad. Sci. U. S. A.* **95**, 4487–4492.
17. Klein CA, Schmidt-Kittler O, Schardt JA, Pantel K, Speicher MR & Riethmuller G (1999) *Proc. Natl. Acad. Sci. U. S. A.* **96**, 4494–4499.
18. Emmert-Buck MR, Bonner RF, Smith PD, *et al.* (1996) *Science*, **274**, 998–1001.
19. Becker I, Becker KF, Rohrl MH & Hofler H (1997) *Histochem. Cell. Biol.* **108**, 447–451.
★ 20. Wang G, Maher E, Brennan C, *et al.* (2004) *Genome Res.* **14**, 2357–2366. – *Original publication describing RCA–RCA.*
21. Lehmann U & Kreipe H (2001) *Methods*, **25**, 409–418.
★ 22. Dean FB, Nelson JR, Giesler TL & Lasken RS (2001) *Genome Res.* **11**, 1095–1099. – *First demonstration of random-primer-based MDA.*
23. Dallman MJ & Porter ACG (1991) *PCR1, a Practical Approach.* Edited by MJ McPherson, P Quirke & GR Taylor. Oxford University Press, Oxford, UK.
24. Liu WH, Kaur M, Wang G, Zhu P, Zhang Y & Makrigiorgos GM (2004) *Cancer Res.* **64**, 2544–2551.
25. Zhang L, Cui X, Schmitt K, Hubert R, Navidi W & Arnheim N (1992) *Proc. Natl. Acad. Sci. U. S. A.* **89**, 5847–5851.
26. Telenius H, Carter NP, Bebb CE, Nordenskjold M, Ponder BA & Tunnacliffe A (1992) *Genomics*, **13**, 718–725.
27. Nelson DL, Ledbetter SA, Corbo L, *et al.* (1989) *Proc. Natl. Acad. Sci. U. S. A.* **86**, 6686–6690.
28. Makrigiorgos GM, Chakrabarti S, Zhang Y, Kaur M & Price BD (2002) *Nat. Biotechnol.* **20**, 936–939.
29. Wang G, Brennan C, Rook M, *et al.* (2004) *Nucleic Acids Res.* **32**, e76.
★ 30. Lizardi PM, Huang X, Zhu Z, Bray-Ward P, Thomas DC & Ward DC (1998) *Nat. Genet.* **19**, 225–232. – *Original demonstration of RCA for mutation detection.*
★ 31. Dean FB, Hosono S, Fang L, *et al.* (2002) *Proc. Natl. Acad. Sci. U. S. A.* **99**, 5261–5266. – *Original description of WGA via MDA.*
★★ 32. Lage JM, Leamon JH, Pejovic T, *et al.* (2003) *Genome Res.* **13**, 294–307. – *Application of MDA to array CGH studies.*
33. Cheung VG & Nelson SF (1996) *Proc. Natl. Acad. Sci. U. S. A.* **93**, 14676–14679.
34. Lovmar L, Fredriksson M, Liljedahl U, Sigurdsson S & Syvanen AC (2003) *Nucleic Acids Res.* **31**, e129.
35. Rook MS, Delach SM, Deyneko G, Worlock A & Wolfe JL (2004) *Am. J. Pathol.* **164**, 23–33.
36. van Gelder RN, von Zastrow ME, Yool A, Dement WC, Barchas JD & Eberwine JH (1990) *Proc. Natl. Acad. Sci. U. S. A* **87**, 1663–1667.

37. Wang E, Miller LD, Ohnmacht GA, Liu ET & Marincola FM (2000) *Nat. Biotechnol.* **18**, 457–459.
38. Lin SL, Chuong CM, Widelitz RB & Ying SY (1999) *Nucleic Acids Res.* **27**, 4585–4589.
39. Sgroi DC, Teng S, Robinson G, LeVangie R, Hudson JR, Jr & Elkahloun AG (1999) *Cancer Res.* **59**, 5656–5661.
40. Hopwood D (1998) In: *PCR3: a Practical Approach*, pp. 1–9. Edited by CS Herrington and JJ O'Leary. Oxford University Press, Oxford, UK.
41. Forozan F, Mahlamaki EH, Monni O, *et al.* (2000) *Cancer Res.* **60**, 4519–4525.

CHAPTER 11

Pre-implantation genetic diagnosis using whole genome amplification

Alan H. Handyside[1], Mark D. Robinson[2] and Francesco Fiorentino[3]

[1]London Bridge Fertility, Gynaecology and Genetics Centre, One St Thomas Street, London Bridge, London SE1 9RY, UK; [2]Leeds Pre-implantation Genetic Diagnosis Centre, Assisted Conception Unit, Leeds General Infirmary, Leeds, UK; [3]Laboratorio Genoma, Rome, Italy

1. INTRODUCTION

Pre-implantation genetic diagnosis (PGD) following assisted conception is now well established clinically as an alternative to conventional pre-natal diagnosis in couples at risk of having children with an inherited disease (1). Controlled ovarian stimulation, egg collection by ultrasound-guided transvaginal needle aspiration, and insemination with the partner's washed sperm provide access to fertilized pre-implantation-stage embryos *in vitro*. Single cells, typically the first and second polar bodies and/or one or two blastomeres, are then removed by micromanipulation from each fertilized zygote or cleavage-stage embryo, respectively, for genetic analysis. This typically involves fluorescent *in situ* hybridization (FISH) and other molecular cytogenetic techniques for detection of chromosomal abnormalities in interphase nuclei, or for detection of single gene defects, PCR-based strategies for DNA amplification, and mutation detection. Finally, unaffected embryos are selected for transfer to the uterus, avoiding the possibility of terminating an affected pregnancy diagnosed at later stages.

The range of genetic defects that can be diagnosed has expanded dramatically since the first births were reported in couples at risk of X-linked conditions and cystic fibrosis (2, 3), and now includes numerical and structural chromosomal abnormalities and most of the common single gene defects (4). The scope of PGD has also been extended to screening for chromosomal aneuploidy in infertile couples (5–7) and for human leukocyte antigen (HLA) typing with or without single gene defect diagnosis with the aim of recovering compatible stem cells from cord blood at birth for transplantation to an existing sick child (8–10). Although precise data are not available, it is now estimated that approaching 1500 babies have been born worldwide following PGD (11).

Whole Genome Amplification: *Methods Express* (S. Hughes and R. Lasken, eds.)
© Scion Publishing Limited, 2005

Diagnosis of single gene defects requires sequence information from both parental chromosomes in the single cell removed from each embryo. This was made possible initially by amplifying a short DNA fragment encompassing the mutation with two rounds of PCR, generally with a nested pair of oligonucleotide primers in the second round (see *Fig. 1A*) (3, 12, 13). With the advent of automated sequencers offering highly sensitive detection and analysis of fluorescent PCR products, strategies were developed for multiplex amplification of several fragments, for example, to combine mutation detection with chromosome-specific sequences to identify the sex of embryos and contamination with exogenous DNA (see *Fig. 1B*) (14, 15). Most recently, this has evolved further to combine multiplex amplification of several short fragments, followed by rapid sequencing or mini-sequencing for sequence/mutation analysis (see *Fig. 1C*) (16, 17).

Highly sensitive amplification strategies, which are capable of detecting as few as one or two target dsDNA molecules in a single cell, are equally highly susceptible to errors through contamination with foreign or previously amplified DNA. One of the advantages of multiplex PCR strategies including chromosome-specific sequences, therefore, is to detect amplification from contaminating exogenous target DNA by detecting markers, often highly polymorphic repeats, not present in the parental chromosomes (see *Table 1*). Another problem when amplifying from single cells is that occasionally one parental allele fails to amplify at random, resulting in allele dropout (ADO). This can also occur by chromosome malsegregation during the early cleavage divisions of the human embryo when only one of the two parental chromosomes has segregated into the cell removed from the embryo for analysis (18). In these situations, multiplex strategies with chromosome-specific markers can identify when ADO occurs and, if the marker is closely linked or intragenic to the gene defect being diagnosed, has the additional advantage of providing a second linkage-based verification of mutation status (19).

2. METHODS AND APPROACHES

2.1. PGD using WGA

The idea of using WGA as a universal first step to enable secondary analysis of a range of sequences without the need to optimize primers and reaction conditions for multiplexing (see *Fig. 1D*) (20) followed the development of the PCR-based primer-extension pre-amplification (PEP) method for haplotyping single sperm (21). PEP has been used for analysis of sex-linked sequences and deletions of the dystrophin gene for PGD of Duchenne muscular dystrophy (22), and to detect a mutation causing familial adenomatous polyposis coli, an autosomal dominant cancer-predisposing syndrome, combined with an intragenic marker (23). However, a number of disadvantages including the limited amplification achieved and inaccuracies in the amplification of highly polymorphic microsatellite repeat sequences, particularly the common dinucleotide repeats, which are valuable as

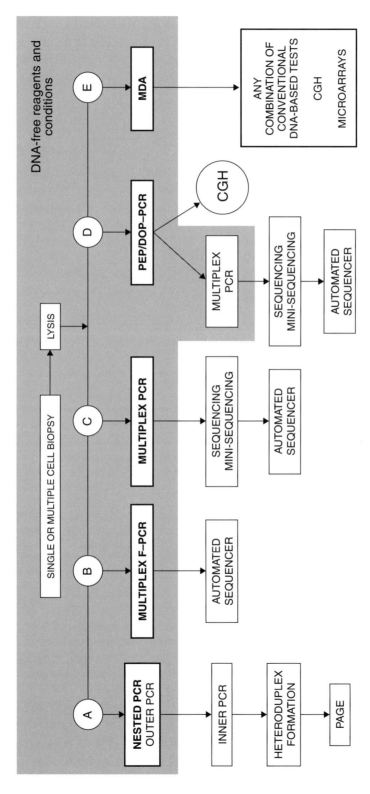

Figure 1. Strategies for amplifying target sequences from single and small numbers of cells for genetic analysis of mutations and other sequences. PAGE, polyacrylamide gel electrophoresis; F–PCR, fluorescent PCR; PEP, primer-extension pre-amplification; DOP, degenerate-oligonucleotide primed; MDA, multiple displacement amplification; CGH, comparative genomic hybridization (see *Table 1* for a more detailed explanation).

Table 1. Comparison of different strategies for amplifying target sequences from single and small numbers of cells for genetic analysis of mutations and other sequences

Method		Feature	Time required (h)*	Advantages	Disadvantages
Amplification of specific target sequences					
A	Nested PCR	Outer PCR can be multiplexed	6	• Increased specificity • Decreased carry-over contamination • Sufficient DNA amplified for conventional analysis	• Repetitive sequences not amplified accurately • Not quantitative
B	Single multiplex fluorescent PCR	Up to eight target fragments	3	• Fast and quantitative • Fingerprinting detects contamination • Screening for major aneuploidies (if informative) can be combined with mutation detection	• Requires carefully optimized set of primers and reaction conditions
C	Multiplex PCR (plus sequencing/mini-sequencing)	Up to 15 target fragments	8	• Moderately fast • Quantitation variable • Screening for major aneuploidies (if informative) can be combined with mutation detection and linked STR markers • Sequencing/mini-sequencing can be applied to any mutation • Reduced ADO	• Limits to multiplexing
WGA					
D	PEP	Linear amplification 80–100 fragments ~800 bp	12	• Multiple fragment analysis without the need for optimization	• Amplification of specific target sequences requires sensitive methods • Repetitive DNA not amplified accurately
	DOP–PCR	Greater quantitative yield Less coverage			
E	MDA	~50 μg DNA product Average 10 kb	10–18	• Universal initial amplification • Sufficient DNA amplified for extensive conventional genetic analysis not requiring specialist facilities	• Variable proportion of human sequence in amplified product • Extensive preferential amplification

*Time required to amplify target sequences prior to analysis by a range of different methods.
STR, short tandem repeat.

linked markers, have discouraged widespread application of this approach (see *Table 1*) (24). Another PCR-based method for WGA, degenerate-oligonucleotide-primed PCR (DOP–PCR), which provides greater amplification but less uniform genome coverage, has been used for comparative genomic hybridization (CGH) and identification of aneuploidy and unbalanced translocations in single cells (6, 7, 24).

The development of WGA using the bacteriophage φ29 DNA polymerase for isothermal multiple displacement amplification (MDA) has several advantages (25, 26). φ29 DNA polymerase has a high processivity, generating amplified fragments of >10 kb by strand displacement, and has proofreading activity resulting in lower misincorporation rates compared with *Taq* DNA polymerase. The random hexamer primers must be thiophosphate modified to protect them from degradation by the 3′→5′ exonuclease proofreading activity of φ29 DNA polymerase (27). Isothermal WGA directly from clinical samples such as blood and buccal swabs has allowed high-throughput genotyping without the need for time-consuming DNA purification steps (28). Sequence representation in the amplified DNA, assessed by multiple single nucleotide polymorphism (SNP) analysis, is equivalent to genomic DNA when amplifying from as little as 0.3 ng of target DNA and amplification bias is superior to PCR-based methods (29).

The principal advantage of MDA for PGD is that sufficient amplified DNA is produced to allow extensive parallel genetic testing and accurate mutation detection by conventional relatively low-sensitivity methods (see *Fig. 1D*) (30, 31). Even from single lymphocytes, the yield of amplified DNA is in the microgram range, allowing, for example, analysis of 20 different loci (including the ΔF508 deletion in exon 10 and two intragenic microsatellite markers in the cystic fibrosis transmembrane conductance regulator (*CFTR*) gene, and nine short tandem repeats used in DNA fingerprinting) by standard, relatively low-sensitivity PCR methods (30). This equals or exceeds the maximum number of loci that have been amplified directly from single cells by multiplex fluorescent PCR, without any need to optimize the conditions for efficient co-amplification (9, 14), and only using a small fraction of the amplified DNA. Furthermore, unlike PCR-based methods (24), the size of all of the polymorphic repeat alleles, including dinucleotide and short tandem repeats, was accurately identified. However, preferential amplification and ADO at heterozygous loci is not eliminated by MDA, and subsequent analysis needs to be carefully optimized and compensated for by increasing the number of loci analysed (see *Table 1*). Alternatively, increasing the number of cells sampled to between two and 20 rapidly reduces these problems (30).

With PCR-based strategies (with or without WGA), separate equipment, isolated clean room facilities, and stringent precautions are essential for the initial stages of amplification to prevent contamination (see *Fig. 1A–D*). This effectively excludes the use of most laboratories where amplification and handling of PCR products on the laboratory bench are commonplace. As a consequence, PGD is a costly, highly specialized procedure only available in a handful of centers with the necessary resources and expertise. In contrast, MDA is easily carried out following embryo biopsy in the DNA-free conditions of clinical embryology laboratories and the products analysed elsewhere by conventional, relatively low-sensitivity approaches (see *Fig. 1E*). By eliminating a significant part of the preliminary work involved in test development by PCR methods alone, this should make it possible to offer PGD for any known genetic defect based on established protocols at a significantly reduced cost. With microgram amounts of MDA product, it may also be feasible to use DNA microarrays:

(i)   to identify karyotype abnormalities by CGH (31);
(ii) to identify haplotypes by large-scale SNP analysis; and
(iii) to screen for a broad range of single gene defects.

A disadvantage of WGA, by either PCR-based or MDA methods, is the significantly increased time involved (see *Table 1*). However, with improvements in embryo culture media, embryo transfer to the uterus is now routinely delayed by 24–30 h following embryo biopsy early on day 3 post-insemination and the successful application of this approach for PGD of β-thalassemia has been reported (32).

To illustrate the power of using MDA for PGD from single or small numbers of cells removed from human embryos, we present here our current methods for cell lysis and MDA, and a combination of protocols that combine testing for:

(i)   mutations causing β-thalassemia by mini-sequencing;
(ii) closely linked short tandem repeat (STR) markers for independent linkage-based verification of mutation status;
(iii) HLA matching using multiple STR markers across the HLA region of chromosome 6; and
(iv) chromosome-specific markers for molecular genetic detection of common aneuploidies.

3. RECOMMENDED PROTOCOLS

Protocol 1

Preparation and lysis of single cells

Equipment and Reagents
- Histopaque-1077 (Sigma)
- Phosphate-buffered saline (PBS) (sterile and calcium/magnesium free)
- PBS with 15 mg/ml polyvinylpyrrolidone (PVP) (high molecular weight; Sigma)
- Alkaline lysis buffer (0.4 M KOH; 10 mM EDTA; 100 mM dithiothreitol)
- Neutralizing buffer (REPLI-g Kit; Qiagen)
- 0.2 ml PCR tubes (DNAse, RNAse and DNA free)
- Mouth pipette with 0.22 μm filter (Millipore) and capillary tube adaptor (Sigma)
- Finely pulled Pasteur pipettes or thick-walled micromanipulator capillary tubing (1 mm outer diameter) (Research Instruments), heat sterilized at 200°C for 2 h
- Mineral oil (embryo culture grade; Sigma)
- 60 mm Tissue culture Petri dishes
- Stereo microscope (Leica MZ12 or equivalent)

Method
1. Separate lymphocytes (and mononuclear cells) from 3 ml of whole blood by centrifugation over Histopaque-1077, according to the manufacturer's instructions[a,b].

2. Carefully remove the buffy coat on the surface of the Histopaque-1077 layer.

3. Wash three times by resuspending cells in 1 ml of PBS (with PVP) and centrifuging at 500 g.

4. Resuspend cells in 1 ml of PBS (with PVP).

5. Place a 10 μl drop of the lymphocyte suspension and a series of 5 μl drops of PBS (with PVP) on a Petri dish and cover drops with 5–7.5 ml mineral oil (sufficient to cover the drops).

6. Transfer a small number of lymphocytes from the lymphocyte suspension drop into the top of one of the PBS (with PVP) drops using a mouth pipette connected to a finely pulled Pasteur pipette or capillary tube.

7. Pick and transfer single cells using a fresh Pasteur pipette or capillary tube, while the lymphocytes remain floating, into 3 μl of PBS in PCR tubes under a stereomicroscope to confirm transfer of the cell.

8. Add 3.5 μl of freshly prepared alkaline lysis buffer to each sample and place the tubes on ice for 10 min to lyse the cells.

9. Stop lysis by adding 3.5 μl of neutralizing buffer and, if not used immediately, store at −20°C.

Notes
[a]Lymphocytes should be prepared in a dedicated laboratory with positive-pressure, high-efficiency particulate air (HEPA) filters taking precautions to avoid contamination.
[b]All sample tubes should be kept in cool racks at approximately ice temperature throughout the procedure.

Protocol 2

WGA using MDA

Equipment and Reagents
- REPLI-g Kit (4× REPLI-g buffer containing exonuclease-resistant, phosphorothioate-modified, random hexamer oligonucleotide primers; REPLI-g DNA polymerase (φ29 DNA polymerase); nuclease-free water; Qiagen)

Method
1. Prepare a master mix of 27 μl of nuclease-free water, 12.5 μl of 4× REPLI-g buffer and 0.5 μl of REPLI-g DNA polymerase[a] for each reaction.

2. Combine the 10 μl of solution from *Protocol 1* with 40 μl of master mix (final reaction volume 50 μl) and mix by gently tapping the tube and centrifuging if necessary.

3. Incubate at 30°C on a thermocycler for 16 h or overnight.

4. Terminate the reaction by raising the temperature to 65°C for 3 min to inactivate the enzyme.

5. Store amplified DNA at 4°C if it is to be used immediately or at –20°C for long-term storage[b,c,d].

Notes
[a]REPLI-g DNA polymerase should be thawed on ice. However, all other components can be thawed at room temperature.

[b]Following amplification, the yield of dsDNA can be measured using PicoGreen reagent (Molecular Probes), following the manufacturer's instructions.

[c]In control reactions without target DNA, amplification still occurs from the primers and gives similar yields.

[d]Due to the presence of amplification in the negative control, if necessary, the proportion of human sequence can be determined using real-time PCR for a chosen target sequence and compared with an unamplified genomic DNA control.

3.1. Downstream applications

3.1.1. Single gene defects

Many different approaches have been used for mutation detection following DNA amplification from single cells (4). However, with the new generation of automated sequencers using capillary electrophoresis, mini-sequencing is being used increasingly because it is universally applicable to mutation detection and can be applied to short amplified fragments, which minimizes ADO (16, 17). Mini-sequencing chemistry is based on the single dideoxynucleotide extension of unlabeled oligonucleotide primers annealing to purified amplified target DNA. Specific mini-sequencing primers, which are exactly one base short of the mutation sites, are used for each mutation under investigation. Primers bind to their complementary templates and *Taq* DNA polymerase then adds a complementary single fluorescent-labeled dideoxynucleoside triphosphate

Protocol 3

Multiplex PCR amplification following WGA

Equipment and Reagents
- GeneAmp PCR System 9700 (Applied Biosystems) (or comparable real-time PCR detection instrument)
- 10× PCR Buffer II (500 mM KCl; 100 mM Tris-HCl (pH 8.3); Applied Biosystems)
- 10× MgCl$_2$ (15 mM)
- dNTPs (25 mM) (Roche Diagnostics)
- AmpliTaq Gold Polymerase (5 units/µl; Applied Biosystems)
- Nuclease-free water (Sigma)

Method
1. Prepare the PCR according to the test being performed. The reaction constituents and cycling conditions for the specific tests are detailed in *Table 2*.

Additional Protocols
The PCR products generated using the profiles described in *Protocol 3* were analysed by either STR genotyping[a] or by mini-sequencing[b]. Both applications were performed using an ABI PRISM 3100 DNA sequencer (Applied Biosystems), according to the manufacturer's protocol.

Notes
[a]Combine 1 µl of each dye-labeled PCR with 0.5 µl of GS500 TAMRA (Applied Biosystems) and 15 µl of Hi-Di formamide (Applied Biosystems) and denature for 4 min at 94°C. Resolve samples by capillary electrophoresis for 30 min on an ABI PRISM 3100 DNA sequencer (Applied Biosystems) and analyze the results using GeneScan Analysis software (Applied Biosystems).

[b]To avoid participation in the mini-sequencing primer-extension reaction, remove primers and unincorporated dNTPs using a Microcon 100 filter (Millipore), according to the manufacturer's protocol. Combine 10 ng of purified PCR product, 5 µl of Ready Reaction Pre-mix (ABI PRISM SNaPshot Multiplex Kit; Applied Biosystems) and 5 pmol of each mini-sequencing primer. Resolve samples by capillary electrophoresis for 15 min on an ABI PRISM 3100 DNA sequencer (Applied Biosystems) using POP-4 polymer (Applied Biosystems).

(ddNTP) at the 3′ end of each primer, according to the sequence. Since the reaction contains only template, primer, and dye-labeled ddNTPs, rather than a mixture with deoxynucleoside phosphates as in a full sequencing protocol, interruption of the reaction occurs after only one incorporation of the dideoxy terminator. This process is repeated in successive rounds of extension and termination. The resulting products, varying in color, can then be analyzed by electrophoresis (see *Fig. 2*). The colors of the final peaks are determined by the specific genotype at the locus under investigation, making it possible to identify the base variation. The mutation site can thus be reliably differentiated between homozygous wild type and mutant (one peak of a specific color; A/green, C/black, G/blue, T/red) or heterozygote. In the latter case, two different-colored peaks occur in the electrophoretogram, one derived from the normal base and the other from the mutated base (see *Fig. 3* in color section).

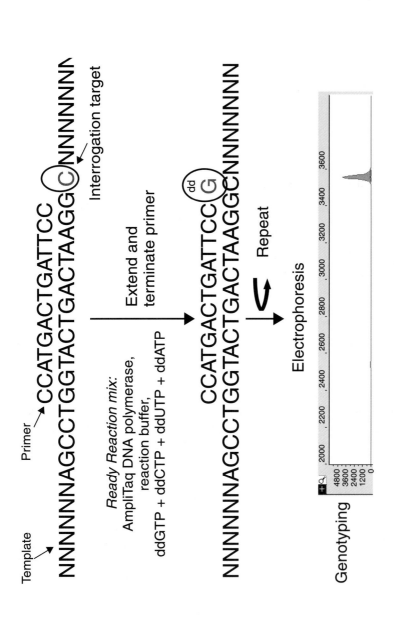

Figure 2. Mini-sequencing single-base extension technique.
Primers bind to their complementary templates and *Taq* DNA polymerase then adds a single fluorescent-labeled dideoxynucleoside triphosphate (ddNTP) to the 3' end of each primer. Since the reaction contains only template, primer, and dye-labeled ddNTPs, and not deoxynucleoside phosphates as in a full sequencing protocol, interruption of the reaction occurs after incorporation of only one of the dideoxy terminators. This process is repeated in successive rounds of extension and termination. The resulting products, varying in color, can then be analyzed by electrophoresis. The mutation site can thus be reliably differentiated from the homozygous wild type, mutant, or heterozygote.

Table 2. PCR constituents for PGD screening

	Informativity testing on individual couples	HLA matching	Aneuploidy screening	Thalassemia and linked STR markers
Genomic DNA	50 ng	1 µl of MDA product (~1 µg)	1 µl of MDA product (~1 µg)	1 µl of MDA product (~1 µg)
10× PCR Buffer II	5 µl (1×)	5 µl (1×)	5 µl (1×)	5 µl (1×)
Concentration of each pair of primers	10 pmol (see *Table 4*)	5 pmol (see *Table 4*)	5 pmol (see *Table 3*)	10 pmol (see *Table 5*)
MgCl$_2$ (mM)	1.5	1.5	1.5	1.5
dNTPs (mM)	200	200	200	200
AmpliTaq Gold polymerase	2.5 units	2.5 units	2.5 units	2.5 units
Ultra-pure water	Up to 50 µl	Up to 50 µl	Up to 50 µl	Up to 50 µl
Cycling conditions	• Initial denaturation of 95°C/10 min • 32 cycles of 95°C/30 s; 60°C/30 s; 72°C/30 s • Final extension of 65°C/60 min	• Initial denaturation of 95°C/10 min • 32 cycles of 95°C/30 s; 55°C/30 s; 72°C/30 s • Final extension of 65°C/60 min	• Initial denaturation of 95°C/10 min • 32 cycles of 95°C/30 s; 55°C/30 s; 72°C/30 s • Final extension of 65°C/60 min	• Initial denaturation of 95°C/10 min • 32 cycles of 95°C/30 s; 55°C/30 s; 72°C/30 s • Final extension of 65°C/60 min

3.1.2. Aneuploidy screening

Pregnancy and live birth rates following *in vitro* fertilization decline rapidly with advancing maternal age. One of the main factors causing this decline is a decrease in egg quality, associated with an increase in errors of female meiosis, particularly meiosis I. As a consequence, a high incidence of aneuploid oocytes and embryos occurs, which are either not viable or develop abnormally and have a high risk of miscarriage. Chromosomal abnormalities also arise during the early cleavage divisions of the fertilized egg as a result of chromosome malsegregation. To screen for these abnormalities and avoid the transfer of aneuploid embryos, embryo biopsy and single-cell analysis by sequential FISH, typically with five to nine chromosome-specific probes, is used for interphase molecular cytogenetic analysis (11, 33, 34). Alternatively, molecular genetic approaches have been used to identify parental chromosomes using STR markers (see *Table 3*) and other markers, both qualitatively and quantitatively (6). In addition, CGH can be used to extend the analysis to the whole karyotype (6, 7, 24). Here, we present a protocol that enables identification of aneuploidy for chromosomes 21, 18, 13, X, and Y using STR markers. Aneuploidy for 21, 18, and 13 are commonly associated with miscarriage or result in an abnormal pregnancy, and sex chromosome aneuploidy is quite common at pre-implantation stages of development. An example showing trisomy 21 is given in *Fig. 4* (in color section).

Table 3. Primers used for screening aneuploidies for chromosomes 21, 18, 13, X, and Y

STR marker	Chromosome	Primer sequences (5′→3′)	Size (bp)	Fluorescent label
D21S11	21	R TGTTGTATTAGTCAATGTTCTCCAG F TCCAGAGACAGACTAATAGGAGGT	200–240	HEX
D21S1414	21	R CCAAGTGAATTGCCTTCTATCTA F GAATAGTGCTGCAATGAACATACAT	190–220	6-FAM
D21S1437	21	R TTCTCTACATATTTACTGCCAACAC F ATATGATGAATGCATAGATGGA	150	HEX
D21S1411	21	R TTGTATTAATGTGTGTCCTTCC F GGAGGCTGAGTCAGGAGAATCA	240–300	TAMRA
D18S386	18	R GCAGGTAGAATCTACGCACCCT F GTACAAACAGCAAACTTTACAGGG	352–370	6-FAM
D18S1002	18	R TGAAGTAGCGGAAGGCTGTAATAT F TCATGTGACAAAAGCCACAC	302–340	TAMRA
D18S535	18	R AGACAGAAATATAGATGAGAATGCA F AGCTGGAGAGGGATAGCATT	150–170	TAMRA
D18S858	18	R TGCATTGCATGAAAGTAGGA F CTGGGCAACAAGAGCAAAACT	200–234	6-FAM
D13S256	13	R GGCCACAGAGGAAGCACATA F GGGACTACCTATGCACACAAAGT	265–290	TAMRA
D13S258	13	R AATGGGATGAGAGAGGAAGACAG F CTGGGCAACAAGAGCAAAACT	172–190	6-FAM
D13S256	13	R GGCCACAGAGGAAGCACATA F TCCATGGATGCAGAATTCACAG	265–280	TAMRA
D13S796	13	R TCTCATCTCCCTGTTTGGTAGC F AAATGCTGGGATCACAGG	180–200	TAMRA
D13S217	13	R CCTGGTGGACTTTTGCTG F GAAGGGAAAATGATGAATAAAACT	200–240	HEX
5′DYS-7	X	R GTCAGAACTTTGTCACCTGTC F GGGCAGTAGCTTTCAGCTTAAAC	156–180	6-FAM
HPRT	X	R CCCTGTCTATGGTCTCGATTCA F CCCTGGGCTCTGTAAAGAATAGTG	150–175	HEX
Amelogenin	X/Y	R ATCAGAGCTTAAACTGGGAAGCTG F GGTGTCTGTGTACAGGTACCTCAG	103–109	6-FAM
DXS6941	X	R GGACCTCCAGAGTTACACATGC F GTGTTACTGGACTCCAGCCTGG	117–135	TAMRA
DXS722	X	R CCTGATCCTGTTCCACTGGG F ACTGGCAACAGAACGAGACTCT	111–138	HEX
DXS1240	X	R AGATCTAGGCAAGGGCAATTAA F TACAACAAGCCAGGTCCTCACT	163–186	6-FAM
DXS1470	X	R GTGTAGTAACTCATATCAAGAGCCG	208–235	HEX

HPRT, hypoxanthine phosphoribosyltransferase; F, forward primer; R, reverse primer; HEX, hexamethylfluorescein; 6-FAM, 6-carboxyfluorescein; TAMRA, 6-carboxytetramethylrhodamine.

3.1.3. HLA matching

For couples who have a child affected by a genetic condition that is treatable by transplantation of HLA-matched hemopoietic stem cells, PGD offers the possibility of combining mutation testing, if the condition is inherited, to avoid the birth of another affected child, with HLA matching. Cord blood stem cells can then be recovered at birth for transfer to the affected child, as stem cells from an HLA-matched sibling donor have the best chance of success. Our approach to HLA matching involves using a number of STR markers across the HLA region of chromosome 6 to ensure that the embryo is HLA compatible and that there has been no recombination (9, 10).

STR haplotyping for family members (father, mother, and affected child) is performed prior to pre-implantation HLA typing, in order to identify the most informative STR markers of the HLA complex to be used in the following clinical PGD cycles. A panel of 50 different STR markers (see *Fig. 5* and *Table 4*) is studied during the set-up phase, to ensure sufficient informativity in all families. For each family, only heterozygous markers presenting alleles not shared by the parents are selected, so that segregation of each allele and discrimination of the four parental HLA haplotypes can be clearly determined. Informativity is also evaluated for STR markers linked to the gene regions involved by mutation, and is thus used to avoid possible misdiagnosis due to the well-known ADO phenomena.

By selecting a consistent number of STR markers evenly spaced throughout the HLA complex, an accurate mapping of the whole region can be achieved. Because genes in the HLA complex are tightly linked and usually inherited in a block, profiles obtained from such markers in father, mother, and affected child allow the determination of specific haplotypes. Thus, the HLA region can be indirectly typed by segregation analysis of the STR alleles and the HLA identity of the embryos matching the affected sibling can be ascertained by evaluating the inheritance of the matching haplotypes. Because segregation of the STR alleles fully corresponds to the direct HLA genotyping, STR haplotyping can be used as a reliable diagnostic tool for indirect HLA matching evaluation. The use of microsatellite markers for this purpose is very useful, since they may provide information on identity over a greater distance within the HLA region compared with classical HLA genes alone, making haplotyping more accurate in predicting compatibility. Another important advantage of using STR markers in pre-implantation HLA matching is that the whole HLA complex can be covered and this allows the detection of recombination events between HLA genes.

An example of the pre-implantation HLA matching procedure using STR haplotyping, in combination with PGD for β-thalassemia, is shown in *Fig. 6*. The strategy presented here enables the selection of HLA-matched embryos and can be performed for any genotype combination, without the need to develop a specific diagnostic design for each couple, as the selected panel of STR markers has already been worked out and can be used for other patients. As a consequence, a substantial shortening of the preliminary phase can be achieved. Recombination, if not detected, could strongly affect the accuracy of the HLA matching procedure. The importance of detecting recombination within the HLA region is demonstrated in *Fig. 7*. Recombination between flanking markers of the paternal or maternal haplotype is detected in two embryos (embryos 1 and 6). In one of them (embryo 1), a single recombination has occurred in the maternal haplotype, between the alleles of the markers D6S105 and MIB. In the other embryo (embryo 6), initially appearing to be HLA matched with the affected sibling, a double recombination event is evident, between markers D6S1683 and

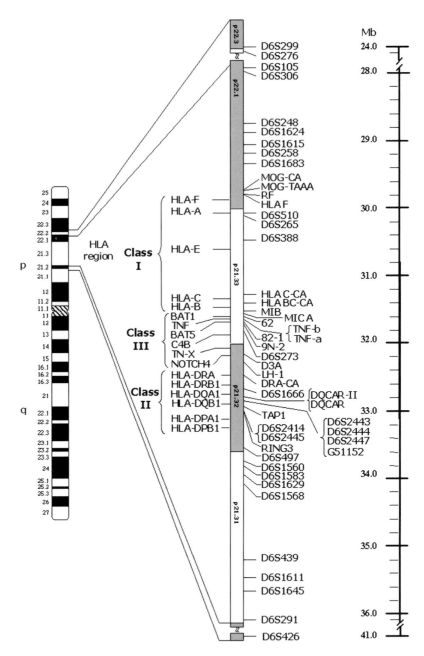

Figure 5. Polymorphic STR markers located throughout the HLA region, on chromosome 6, used in the pre-implantation HLA matching procedure.
The STR markers are ordered from the telomere (top) to the centromere (bottom) and their position is compared with genes of the HLA complex. D6S299, D6S276, and D6S426 markers are located outside the HLA region. All STR markers are dinucleotide repeats, except for RF, which is a trinucleotide repeat, and MOG-TAAA, D6S2414, D6S2415, and D6S497, which are tetranucleotide repeats. Mb, megabases.

Table 4. Oligonucleotide primer sequences of selected STR markers located in the HLA complex used for pre-implantation HLA matching

STR locus	Primer sequences (5′→3′)	Size range (bp)	Fluorescent label
D6S299	F CCATGCAGTAACTCAGATCTAGGA R ATGTCTCTCTTCTCTCCCTCCC	160–176	HEX
D6S306	F AAGGTTTGTCAAACATCCCATC R GGTTTGAGAGTTTCAGTGAGCC	155–160	6-FAM
D6S1615	F AATTCCTCTGCTCTCTGGGATT R AGCCTGGGTAACAGAGCAAGA	98–122	6-FAM
D6S258	F GCAAATCAAGAATGTAATTCCC R GCTTTGTAGTTCTTTTTGTGGA	121–131	HEX
D6S1683	F ACATGTATCCGAGAACTTAAAGT R AAGTAGAGACAGGATTTCTTGT	170–178	6-FAM
MOG-TAAA	F GTGGGCACCTATAATACCAGCTAC R GGGTTAGAAGTGTGCTTATGAA	215–227	TAMRA
HLA-F	F TATGCTCAGGTACAACTTTTCCAG R TGAACTTGTCCTGAGAATGAAGG	260–275	6-FAM
D6S2971	F CTGTCCTATTTCATATGCTCAGGTA R TTGTCCTGAGAATGAAGGTCTAGA	230–263	6-FAM
D6S388	F GCTGATGGAGAATGAAATATGG R GGTTAGACGTAGCTTAAGAGAGAAT	150–155	TAMRA
D6S1666	F GTTGGGCAGCATTTGTAGATTTC R ACCCAGCATTTTGGAGTTGTGT	112–142	HEX
D6S2443	F CCATACCAAAGTAAAACCCAGTG R CATTTGATACTGAGGATGAAGGG	180–188	6-FAM
D6S2444	F GGGAGCATTTGTGTATTTCTGTATG R AATGATTCATGAGCCAAGAACC	137–145	HEX
D6S2414	F AACTGGGCTGAGATGTACCACT R GACTCAAGGAGAGGAATGTGTG	155–165	TAMRA
D6S2415	F CAGCCCTTAACAGCTTTATTGG R ATGAACCTGACTGTGGTGATGA	152–157	HEX
D6S497	F CCTGGGCAACAAGAGTGAACT R TTGGCTGTTGAATTGTGAGAGT	129–140	6-FAM
D6S1560	F TCCTTGGTGGTAGTGTTTCTAA R TGAGTCAAGTGAGAAACAGAGAG	130–146	6-FAM
D6S1583	F CCCTAACCTGCTTCTACTGATCA R CTCAGGGACAGACAACCTCTG	130–138	TAMRA
D6S1629	F CACAGTGACTTGTACTGAAAGCTCA R GGCTCCCCAATTATCTCTGC	155–165	HEX
D6S1568	F AGATATCCCCACCAAGGCAG R AGCTAGGCCAGGCCGTGT	127–152	6-FAM
D6S1611	F GGATTTCTTGCAAAACAAACCC R AAGGGCTGAGTTTCTTCTTGGG	180–185	HEX
D6S1645	F ACAGAGTGAGACTCTGTCGCAAAC R CCCACTTAGCAGACAGAGAGATAGA	160–167	6-FAM
D6S276	F TCAATCAAATCATCCCCAGAAG R GGGTGCAACTTGTTCCTCCT	190–220	HEX
D6S291	F GTCTAAAATATCCATCCGGCAT R TTAATTGTGGTGATGGTTTCAC	156–166	6-FAM
D6S426	F ACTCCCCCAAAAATGTAGTCAT R AAAATGCACGTACCTAGTCCTC	112–130	NED
D6S273	F TGAGTATTTCTGCAACTTTTCTGTC R AAACCAAACTTCAAATTTTCGG	135–146	HEX
D6S265	F TCGTACCCATTAACCTACCTCTCT R TCGAGGTAAACAGCAGAAAGATAG	110–125	6-FAM

Table 4. Continued

STR locus	Primer sequences (5′→3′)	Size range (bp)	Fluorescent label
MICA	F GAAAGTGCTGGTGCTTCAGAGTC R CTTACCATCTCCAGAAACTGCC	170–180	NED
TNF-α	F GCCTCTAGATTTCATCCAGCCAC R CCTCTCTCCCCTGCAACACA	97–121	6-FAM
TAP1CA	F TCATACATCTGCTTTGATCTCCC R GGACAATATTTTGCTCCTGAGG	195–215	NED
TNF-β	F TGTGTTGCAGGGGAGAGAGG R GGCCACAGAGCAAGACACCA	100–118	HEX
D6S2447	F CTGCATTTCTCTTCCTTATCACTTC R TTTGAGAGGTGTGCATGTTACC	180–202	6-FAM
D6S510	F TTTGTCTTTCCCAATGTACTACAC R GCTACTACTTCACACCAATTAGGA	140–155	HEX
D3A	F CATCCATGACAGAAAGCAGAGC R CCTGCCTTCTGTAAGCCTCAG	181–215	NED
62	F GATTTCATCCAGCCACAGGA R TCCAATCACCTCTGCTCACC	140–170	HEX
82-1	F GAGCCAGGATGGAGACCAAA R CCTGGATAACAGAACGAGACCC	100–122	6-FAM
G51152	F GGAAAAGAGCTCACGCACAT R CCTGCCATCATGACTTCAAG	145–158	NED
LH-1	F GCTAGTCTGTGCCAAGGAACTC R ACCTTACTGGGCACAAATTCAC	126–160	6-FAM
Ring3CA	F GCCGCAGTTTAACTGTTCCTT R GAAATGTTAGGTCAGAACCACAGA	124–130	HEX
MOG-CA	F AGATCACCTCGAGTGAGTCTCTT R TTGACCATGGGTAACTGAAGC	205–235	6-FAM
DRA-CA	F ACTTTCCTAATTCTCCTCCTTC R GCATGAGTAAACTATGGAATCTC	122–140	HEX
D6S439	F CCCCTATTCTCCACCCACTAGA R CAGCCTCAGGGAAGACACATT	116–130	NED
DQCAR	F CTGCATTTCTCTTCCTTATCACTTC R TGGCCAATCAGAATCTTTCCTA	150–175	6-FAM
9N-2	F TGGGTAACAGAGCAAGACTCTGT R TGGGATTGCAGATGTGTTACAC	100–110	HEX
MIB	F CGTTTTCAGCCTGCTAGCTTAT R CCACAGTCTCTATCAGTCCAGATTC	155–186	HEX
D6S105	F AACAAGAGCAAAACTCCGTCTC R TCACCTTGATATCTTATTACCCTGG	141–155	NED
DQCARII	F TTGGGCAGCATTTGTAGATTTC R GCAAGAATCCAGCATTTTGG	118–138	6-FAM
HLABC-CA	F GTCAAGCATATCTGCCATTTGG R ACTTGGGCAATGAGTCCTATGA	113–144	HEX
HLAC-CA	F CGGCAAGAGACTCTGATGAGAA R GTAGCTGGGATTACAGGTGCCT	156–173	NED
D6S248	F CGAGATCAAGCCACTGCACT R CAGGAATGGTGAGAAGGGAAA	151–165	HEX
D6S1624	F TATAACCCCAGGTGTTTGTGG R GGAAGTCTTCAGTGGAGAGAGTG	200–220	6-FAM

MOG-TAAA, myelin oligodendrocyte glycoprotein (TAAA)$_n$ repeat; HLA-F, human leukocyte antigen F; MICA, MHC class I polypeptide-related sequence A; TNF-α, tumor necrosis factor-α; TAP1CA, transporter associated with antigen processing 1CA; TNF-β, tumor necrosis factor-β; Ring3CA, bromodomain-containing protein 2 (CA)$_n$ repeat; MOG-CA; myelin oligodendrocyte glycoprotein (CA)$_n$ repeat; DRA-CA, HLA class II histocompatibility antigen, DR-α chain precursor (HLA-DRA) (CA)$_n$ repeat; MIB, D6S2810, 24.9 kb centromeric of HLA-B; F, forward primer; R, reverse primer; HEX, hexamethylfluorescein; 6-FAM, 6-carboxyfluorescein; TAMRA, 6-carboxytetramethylrhodamine.

D6S265. This occurrence, which was only detected by using a consistent number of STR markers able to determine a fine mapping of the whole HLA region, if missed, could lead to an HLA-genotyping misdiagnosis, and the embryo would be erroneously diagnosed as HLA identical. Hence, the reliability of the procedure is

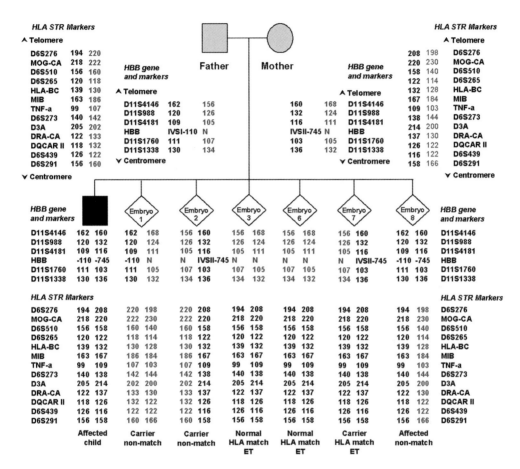

Figure 6. Pre-implantation HLA matching in combination with PGD for β-thalassemia, resulting in the birth of two twins, both HLA matched with the affected sibling.
Specific haplotypes were determined by genomic DNA analysis of HLA STR markers and hemoglobin-β (HBB)-linked markers from father, mother (upper panel), and affected child (lower panel, left side, black square). Informative STR markers are ordered from the telomere (top) to the centromere (bottom). The numbers next to the STR markers represent the size of PCR products in bp. Paternally and maternally derived HLA haplotypes matched to the affected child are shown in bold. STR alleles linked to the paternal and maternal mutations are also shown in bold. Examples of different results of HBB mutation analysis and HLA haplotyping from biopsied blastomeres are shown in the lower panel. Paternally and maternally derived haplotypes from each embryo are shown on the left and the right, respectively. The HLA identity of the embryos with the affected sibling has been ascertained by evaluating the inheritance of the matching haplotypes. Embryos 1, 2 (carriers), and 8 (affected) represent HLA-non-identical embryos. Embryos 3 and 6 were diagnosed as normal, and embryo 7 as a carrier, and were HLA matched with the affected sibling and transferred, resulting in the birth of HLA-matched unaffected twins (the babies originated from embryos 3 and 6). ET, embryo transfer.

strongly correlated with the number of STR markers used for HLA haplotyping. Furthermore, the combined use of a multiplex HLA STR marker system has allowed the detection of aneuploidies of chromosome 6. The relevance of aneuploidy testing for chromosome 6 is seen in *Fig. 7*. One of the embryos tested in this case has only one maternal chromosome 6 (embryo 5), and one (embryo 13) has an extra maternal chromosome, consistent with a diagnosis of monosomy 6 and trisomy 6, respectively, making them unacceptable for transfer. The

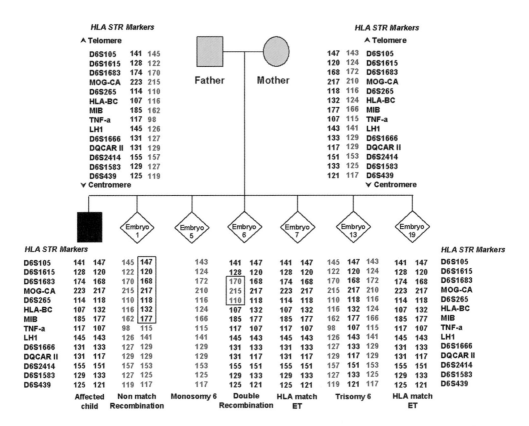

Figure 7. Avoidance of pre-implantation HLA matching misdiagnosis due to recombination and aneuploidy.
The upper panel shows determination of the different haplotypes from father, mother, and affected child (lower panel, left side, black square) by segregation analysis of the alleles obtained after STR genotyping of the HLA region. Informative STR markers used are ordered from the telomere (top) to the centromere (bottom). Paternally and maternally derived HLA haplotypes, matched to the affected child, are shown in bold. Examples of different results of HLA haplotyping from biopsied blastomeres are shown in the lower panel. Embryo 5 has no paternal chromosome present (monosomy 6); embryo 13 shows an extra maternal chromosome (trisomy 6); embryo 1 shows a single recombination event in maternal haplotypes between the alleles of the markers D6S105 and MIB (boxed); in embryo 6, initially appearing to be HLA matched with the affected sibling, a double recombination event was observed between the markers D6S1683 and D6S265 (boxed). Embryos 7 and 19 were diagnosed as HLA matched and were transferred. ET, embryo transfer; HBB, hemoglobin-β; MIB, D6S2810, 24.9 kb centromeric of HLA-B; MOG, myelin oligodendrocyte glycoprotein; TNF-α, tumor necrosis factor-α.

oligonucleotide primer sequences of selected STR markers located in the HLA complex that are used for pre-implantation HLA matching are shown in *Table 4.*

3.1.4. PGD of β-thalassemia combined with HLA matching

The protocol involves the amplification of the gene regions affected by mutations and a panel of informative STR markers, selected during the preclinical work-up, linked to these regions for ADO detection (see *Fig. 8* and primer list in *Table 5*). One of the primers for each microsatellite is labeled with a fluorescent dye (e.g. 6-Fam, Hex, Ned) so that it could be visualized on an automated DNA Sequencer (ABI PRISM 3100; Applied Biosystems). STR markers with overlapping size ranges were labeled with different fluorochromes in order to analyze them in the same capillary electrophoresis run.

Table 5. Primers used for amplification of the hemoglobin-β (HBB) gene and linked STR markers for ADO detection

Gene region/marker	Primer sequences (5′→3′)	Size (bp)	Fluorescent label
HBB gene exon 1	F CATCACTTAGACCTCACCCTGT R TCTCCTTAAACCTGTCTTGTAACC	303	
HBB gene exon 2	F TGGGTTTCTCATAGGCACTGA R AAAGAAAACATCAAGGGTCCC	398	
HBB gene exon 3	F TATCATGCCTCTTTGCACCATT R CAGTTTAGTAGTTGGACTTAGGGAA	449	
HBB gene IVSII	F CCTAATCTCTTTCTTTCAGGGCAAT R GGTATGAACATGATTAGCAAAAGGG	270	
TH01	F GTGGGCTGAAAAGCTCCCGATTAT R GTGATTCCCATTGGCCTGTTCCTC	156–178	6-FAM
D11S4146	F GGTTAAGCAGAGTTTAATAGGC R CTACCAAACATGATTCCTAGGA	163–180	HEX
D11S988	F CACAGAAAATAGTTCAGACCACCAT R TGGGACAAGAGAAAGTTGAACATAC	127–148	HEX
D11S4181	F CTGGGCAACAAGAGTAAGTCTCT R CCTTAAGAACTGAGACCAAGAACA	117–135	6-FAM
B-STR	F AGACTGGAGTAAAGGAAATGG R GATGCCACAGCAGGTG	100–118	6-FAM
D11S1760	F ATCTCAAGTGTTTCCCCACAAC R CTGCATCATGACTTGAAAAACG	106–125	6-FAM
D11S1338	F CCACACAGATTCACTTAAAGCAA R GCTACTTATTTGGAGTGTGAATTTC	135–148	HEX
D11S1997	F TTCCTAAGAAAGATAAAGCACCAG R CAATTGACAGTGGATTTTTGAC	143–150	6-FAM
D11S1331	F GATGTTTAGATGCACAAGACACAGA R CTTCCTTCGTCTTTCTCACTTTTAC	156–178	HEX
D11S4149	F GGCTAAAAAGGCAACAGATAACATC R CCATATATAGAATCACACTGGCCAA	161–180	6-FAM

F, forward primer; R, reverse primer; HEX, hexamethylfluorescein; 6-FAM, 6-carboxyfluorescein.

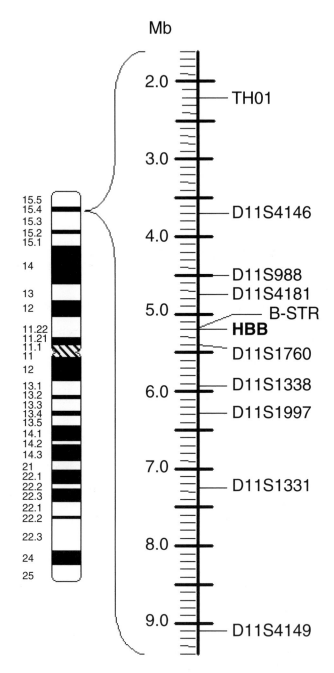

Figure 8. Polymorphic STR markers linked to the hemoglobin-β (HBB) gene on chromosome 11.
STR markers are ordered from the telomere (top) to the centromere (bottom). Mb, mega bases.

4. REFERENCES

1. Sermon K, van Steirteghem A & Liebaers I (2004) *Lancet*, **363**, 1633–1641.
2. Handyside AH, Kontogianni EH, Hardy K & Winston RM (1990) *Nature*, **344**, 768–770.
3. Handyside AH, Lesko JG, Tarin JJ, Winston RM & Hughes MR (1992) *N. Engl. J. Med.* **327**, 905–909.
4. Thornhill AR & Snow K (2002) *J. Mol. Diagn.* **4**, 11–29.
5. Gianaroli L, Magli MC, Ferraretti AP & Munne S (1999) *Fertil. Steril.* **72**, 837–844.
6. Wells D, Escudero T, Levy B, Hirschhorn K, Delhanty JD & Munne S (2002) *Fertil. Steril.* **78**, 543–549.
7. Wilton L, Williamson R, McBain J, Edgar D & Voullaire L (2001) *N. Engl. J. Med.* **345**, 1537–1541.
8. Verlinsky Y, Rechitsky S, Schoolcraft W, Strom C & Kuliev A (2001) *JAMA*, **285**, 3130–3133.
★ 9. Fiorentino F, Biricik A, Karadayi H, *et al.* (2004) *Mol. Hum. Reprod.* **10**, 445–460. – *Description of PGD of single gene defects combined with HLA matching.*
10. Fiorentino F, Kahraman S, Karadayi H, *et al.* (2005) *Eur. J. Hum. Genet.* (in press).
11. Verlinsky Y & Kuliev A (2003) *Reprod. Biomed. Online*, **7**, 145–150.
12. Coutelle C, Williams C, Handyside A, Hardy K, Winston R & Williamson R (1989) *BMJ*, **299**, 22–24.
13. Holding C & Monk M (1989) *Lancet*, **2**, 532–535.
14. Findlay I, Matthews PL, Mulcahy BK & Mitchelson K (2001) *Mol. Cell. Endocrinol.* **183** (Suppl. 1), S5–12.
15. Findlay I, Quirke P, Hall J & Rutherford A (1996) *J. Assist. Reprod. Genet.* **13**, 96–103.
16. Fiorentino F, Magli MC, Podini D, *et al.* (2003) *Mol. Hum. Reprod.* **9**, 399–410.
17. Bermudez MG, Piyamongkol W, Tomaz S, Dudman E, Sherlock JK & Wells D (2003) *Prenat. Diagn.* **23**, 669–677.
18. Kuo HC, Ogilvie CM & Handyside AH (1998) *J. Assist. Reprod. Genet.* **15**, 276–280.
19. Lewis CM, Pinel T, Whittaker JC & Handyside AH (2001) *Hum. Reprod.* **16**, 43–50.
20. Snabes MC, Chong SS, Subramanian SB, Kristjansson K, DiSepio D & Hughes MR (1994) *Proc. Natl. Acad. Sci. U. S. A.* **91**, 6181–6185.
21. Zhang L, Cui X, Schmitt K, Hubert R, Navidi W & Arnheim N (1992) *Proc. Natl. Acad. Sci. U. S. A.* **89**, 5847–5851.
22. Kristjansson K, Chong SS, van den Veyver IB, Subramanian S, Snabes MC & Hughes MR (1994) *Nat. Genet.* **6**, 19–23.
23. Ao A, Wells D, Handyside AH, Winston RM & Delhanty JD (1998) *J. Assist. Reprod. Genet.* **15**, 140–144.
24. Wells D, Sherlock JK, Handyside AH & Delhanty JD (1999) *Nucleic Acids Res.* **27**, 1214–1218.
25. Dean FB, Hosono S, Fang L, *et al.* (2002) *Proc. Natl. Acad. Sci. U. S. A.* **99**, 5261–5266.
★★ 26. Lasken RS & Egholm M (2003) *Trends Biotechnol.* **21**, 531–535. – *Review of MDA applications and its impact on numerous scientific fields.*
27. Dean FB, Nelson JR, Giesler TL & Lasken RS (2001) *Genome Res.* **11**, 1095–1099.
28. Hosono S, Faruqi AF, Dean FB, *et al.* (2003) *Genome Res.* **13**, 954–964.
29. Lovmar L, Fredriksson M, Liljedahl U, Sigurdsson S & Syvanen AC (2003) *Nucleic Acids Res.* **31**, e129.
★ 30. Handyside AH, Robinson MD, Simpson RJ, *et al.* (2004) *Mol. Hum. Reprod.* **10**, 767–772. – *First demonstration of MDA from single and small numbers of cells for extensive secondary genetic analysis of a range of genetic markers and potential application for PGD.*
31. Hellani A, Coskun S, Benkhalifa M, *et al.* (2004) *Mol. Hum. Reprod.* **10**, 847–852.
32. Hellani A, Coskun S, Tbakhi A & Al-Hassan S (2005) *Reprod. Biomed. Online*, **10**, 376–380.
33. Munne S, Sandalinas M, Escudero T, *et al.* (2003) *Reprod. Biomed. Online*, **7**, 91–97.
34. Verlinsky Y & Kuliev A (2003) *Fertil. Steril.* **80**, 869–70.

APPENDIX 1
List of suppliers

ABgene – www.abgene.com
Alexis Corporation – www.alexis-corp.com
Amersham Biosciences – www.amershambiosciences.com
Anachem Ltd – www.anachem.co.uk
Appleton Woods Ltd – www.appletonwoods.co.uk
Applied Biosystems – www.appliedbiosystems.com
Arcturus Engineering – www.arctur.com
AutoGen, Inc. – www.autogen.com
Axon Instruments – www.axon.com

Beckman Coulter, Inc. – www.beckman.com
Becton, Dickinson and Company – www.bd.com
Bio-Rad Laboratories, Inc. – www.bio-rad.com
BOC Group – www.boc.com
Brosch direct Ltd – www.broschdirect.com

Calbiochem – www.calbiochemicom
Cambridge Scientific Products – www.cambridgescientific.com
Carl Zeiss – www.zeiss.com
Chemicon International, Inc. – www.chemicon.com
Corning, Inc. – www.corning.com

DakoCytomation – www.dakocytomation.com
Difco Laboratories – www.difco.com
Dionex Corporation – www.dionex.com
DuPont – www.dupont.com

Elliot Scientific Ltd – www.elliotscientific.com
European Collection of Animal Cell Culture – www.ecacc.org.uk

Fermentas – www.fermentas.com
Findel Education Ltd – www.fipd.co.uk
Fisher Scientific International – www.fishersci.com
Fluka – www.sigma-aldrich.com
Fluorochem – www.fluorochem.co.uk

Gibco BRL – www.invitrogen.com
Goodfellow Cambridge Ltd – www.goodfellow.com
Greiner Bio-One – www.gbo.com

Harlan – www.harlan.com
Hybaid – www.hybaid.com
HyClone Laboratories – www.hyclone.com

ICN Biomedicals, Inc. – www.icnbiomed.com
Insight Biotechnology – www.insightbio.com
Invitrogen Corporation – www.invitrogen.com

Jencons-PLS – www.jencons.co.uk

Kendro Laboratory Products – www.kendro.com
Kodak: Eastman Fine Chemicals – www.eastman.com

Lab-Plant Ltd – www.labplant.com
Lancaster – www.lancastersynthesis.com
Leica – www.leica.com
Life Technologies Inc. – www.lifetech.com
LOT-Oriel – www.lot-oriel.com

Merck, Sharp and Dohme – www.msd.com
MetaChem – www.metachem.com
Millipore Corporation – www.millipore.com
Miltenyi Biotec – www.miltenyibiotec.com
Molecular Machines and Industries – www.molecular-machines.com
MWG Biotech – www.mwg-biotech.com

National Diagnostics – www.nationaldiagnostics.com
New England BioLabs, Inc. – www.neb.com
Nikon Corporation – www.nikon.com

Olympus Corporation – www.olympus-global.com
Optivision Ltd – optivision.co.uk

PALM Microlaser Technologies AG – www.palm-microlaser.com
Perbio Science – www.perbio.com
PerkinElmer, Inc. – www.perkinelmer.com
Pharmacia Biotech Europe – www.biochrom.co.uk
Photonic Solutions plc – www.psplc.com
Promega Corporation – www.promega.com

Qiagen N.V. – www.qiagen.com

R&D Systems – www.rndsystems.com
Roche Diagnostics Corporation – www.roche-applied-science.com

Sanyo Gallenkamp – www.sanyogallenkamp.com
Sarstedt – www.sarstedt.com
Schleicher and Schuell Bioscience, Inc. – www.schleicher-schuell.com
Scientifica – www.scientifica.uk.com
Serotec – www.serotec.com
Shandon Scientific Ltd – www.shandon.com
Sigma-Aldrich Company Ltd – www.sigma-aldrich.com
Sorvall – www.sorvall.com
Stratagene Corporation – www.stratagene.com

Takara Bio Inc. – www.takara-bio.com
Thames Restek – www.thamesrestek.co.uk
Thermo Electron Corporation – www.thermo.com
Thistle Scientific – www.thistlescientific.co.uk

Vector Laboratories – www.vectorlabs.com
VWR International Ltd – www.bdh.com

Wolf Laboratories – www.wolflabs.co.uk

York Glassware Services Ltd – www.ygs.net

Index

adaptor
    ligation, 35, 48, 50
    molecule, 54
    sequences, 41
adaptor-ligation PCR of randomly sheared
    genomic DNA (PRSG), 7, 48, 49, 51,
    53–55, 57
adenomatous polyposis of the colon (APC),
    51, 55
Affymetrix, 72
allele dropout (ADO), 112, 117, 164, 167,
    170, 175, 181
amplification
    bias, 6, 20, 47, 99, 100, 108, 111, 113,
      126, 167
    fidelity, 78, 81, 93, 97
amplified fragments, 14
anaerobes, 132
aneuploidies, 168
antisense RNA (aRNA), 83, 87, 89, 92, 94,
    97
archaebacteria, 122
archeological, 115
archival tissue, 33, 60, 159
array CGH, 33–35, 41, 44, 55–57, 72, 73,
    117, 151, 153, 155–158
array platform, 41
array(s), 111, 127, 132, 144
association studies, 112
autoradiography, 103, 104

bacterial
    artificial chromosome (BAC), 48, 107
    artificial chromosome (BAC) arrays, 33
    genome, 143
balanced PCR, 151
BigDye terminator, 16, 69, 70

biofilm, 120
blastomere, 163, 179, 180
blood, 63, 99, 107, 108, 113, 117, 168
blunt-ended, 51, 82
blunt ends, 81
breast cancer, 34, 149
buccal cells, 59
buccal swab, 1, 60, 63, 107–109, 113
buffy coat(s), 63, 107, 108, 169

cancer, 1, 30, 112, 116, 149, 157, 164
capillary electrophoresis, 30
carcinogenesis, 24
cDNA, 4, 7, 151–153, 157–159
    arrays, 33, 44
cell morphology, 132
cell sorting, 1
centromere, 176, 179, 180, 182
ChIP–chip, 77–79, 81, 82, 84, 85, 93, 94
chromatin immunoprecipitation (ChIP), 3,
    7, 77, 78, 81–85, 94, 95, 97
chromosomal
    abnormalities, 163
    CGH, 33–35, 39, 40, 43
    deletions, 26
    instability, 29
    translocations, 100
chromosome(s), 6, 12, 23, 40, 61, 113, 138,
    142, 143, 156–158, 164, 168,
    173–176, 180, 182
clonal expansion, 24
colon cancer, 25
colorectal, 74, 117
colorectal cancer(s), 29, 30
comparative genomic hybridization (CGH),
    1, 7, 11, 12, 23, 33, 34, 44, 47, 48,
    57, 60, 72–75, 117, 165, 168

comparative genomic(s), 142, 146
copy number, 72, 117, 128, 155
Cy3, 35, 72
Cy5, 35, 72
cytogenetic, 163

degenerate-oligonucleotide-primed (DOP)
    PCR, 2, 7, 11–15, 17–21, 23-25, 33,
    47, 100, 150, 165–167
degenerate primer(s), 2, 11, 18, 23, 100
dinucleotide, 30, 71, 167, 176
diploid, 144
dissected cells, 54
dissociation constant, 104
DNA
    degradation, 63, 75, 154
    fragment(s), 7, 12, 14, 49, 51, 53, 60,
        144, 150, 164
    fragmentation, 3, 78, 84, 96, 159
    replication, 6
    sequencer, 16
    yield(s), 15, 27, 29, 38, 43, 68, 110
double helix, 100
Drosophila, 18

ecology, 142
embryo(s), 164, 168, 173
endonuclease, 100
endonucleolytic cleavage, 3
Escherichia coli, 102, 119, 121, 122, 123,
    126, 127, 128, 129
esophageal carcinoma, 55
eukaryotic, 48, 144
exonuclease, 4, 24, 96, 106, 107, 109, 150,
    152, 153, 159, 167, 170
extremophiles, 132, 142

fingerprints, 99, 107
fixed tissue(s), 38, 44, 47, 60, 63, 66, 68,
    72, 75
flow cytometer, 119
flow cytometry, 121
flow sorted, 12, 23, 124
fluorescence-activated cell sorting (FACS),
    25, 122, 126
fluorescence-based micosatellite analysis,
    31
fluorescence correlation spectroscopy
    (FCS), 12, 14, 17–21
fluorescent dyes, 78

fluorescent in situ hybridization (FISH), 34,
    119, 127, 132, 133, 136, 137, 139,
    140, 144, 163, 173
fluoresence, 82
forensic, 61, 115
formalin fixation, 150, 154
formalin-fixed paraffin-embedded (FFPE) ,
    13, 33, 35, 44, 48, 55, 149–152,
    154, 155, 157–159
fragmented DNA, 6, 52, 57, 62, 123, 151,
    152
fresh tissue, 60, 63, 66, 67, 75, 151
frozen tissue, 43, 48

gene expression, 149
gene rearrangements, 100
genetic
    alterations, 1, 47
    amplifications, 155
    analysis, 163, 165, 166
    defect, 163
    diversity, 124
    instability, 29
    profiles, 149
genome
    arrays, 94
    coverage, 108
    representation(s), 34, 48, 55
GenomePlex, 7, 59, 60, 62–67, 69–73,
    75
genomic
    amplifications, 33
    analysis, 1, 33
    deletions, 33
    diversity, 143
    DNA(s), 1, 2, 6, 11, 12, 14, 15, 18–21,
        33–39, 41, 42, 44, 48, 55, 59,
        60, 75, 78, 79, 83, 84, 95, 97,
        99, 100, 102, 105, 106, 108,
        110–114, 119, 120, 122,
        124–126, 128, 130, 132, 141,
        143–146, 149, 151, 153–159,
        167, 173, 179
    sequencing, 3
GenomiPhi, 19–21, 153
genotype(s)/genotyping, 12, 18–21, 25, 33,
    34, 47, 55, 72, 99, 100, 110–113,
    115–117, 119, 120, 167, 171, 175,
    179, 180
geochemistry, 142

haplotype(s), 168, 175, 179, 180
heterogeneous, 149, 150
heterozygote, 114, 167
heterozygous, 30, 111–114, 117, 171, 175
high fidelity, 100, 105, 110
high molecular weight, 47, 49, 60, 66, 81, 99, 100, 108, 113, 115, 119, 154
high stringency, 11, 12
high throughput, 47, 60, 77, 99, 120, 167
histological, 26, 59
homogeneous, 150
homozygote, 55
homozygous, 111–114, 117, 171
horizontal gene transfer, 124
human genome, 11, 108
human leukocyte antigen (HLA), 163, 168, 173–177, 179–181
hydrodynamic shear/shearing, 48–50, 55, 57
hyper branched, 5, 102, 150
hyper branching, 4, 101, 102, 104, 115

immunoprecipitated, 78
improved (I)PEP–PCR, 2, 7, 23–25, 27–32
in situ hybridization (ISH), 12, 23
in vitro
    DNA amplification, 4
    fertilization (IVF), 132, 173
    transcription (IVT), 79–84, 86–93, 95, 96, 151
inflammatory cells, 26
interphase, 163, 173
interspecies, 146
interspersed repetitive sequence (IRS) PCR, 2
intragenic, 164, 167
intraspecies, 146
I-PEP, 25, 28–30, 32, 100, 111
isothermal, 80, 87, 102, 132, 139, 150, 167

karyotype, 168, 173

laser capture microdissection (LCM), 1, 37, 39, 48, 54, 61, 116, 117, 150
laser microdissection, 24, 26, 30, 32, 54
leukocytes, 149
ligation-mediated PCR (LM–PCR), 35, 77
linear amplification, 3, 7, 79, 84, 85, 87, 93, 94, 166
linkage groups, 142, 144

long PCR, 24
loss of heterozygosity (LOH), 26, 30, 31, 34, 35, 112, 151
low molecular weight, 47, 66, 81, 92, 93, 95, 96
low stringency, 11, 12, 23
lymphocytes, 167, 169

melanoma, 34
metagenomic(s), 119, 120, 130, 131, 143, 144
metaphase
    CGH, 40
    chromosome, 40
methanol-fixed paraffin-embedded (MFPE), 48, 54, 55
microarray(s), 1, 4, 33, 72, 77, 90, 93, 96–98, 132, 144, 149–151, 157, 165, 168
microbe(s), 124, 133, 143
microbial, 119, 120, 132, 133, 142, 143, 146
microcapillary, 133, 135, 141
microdissection, 12, 23, 25, 26, 30, 33–37, 47, 53, 54, 150
microinjector(s), 132, 133, 141
micromanipulation, 119, 121, 127, 132, 133, 141, 144, 146, 163
micromanipulator, 134, 136, 141
microsatellite(s), 1, 7, 11, 12, 18–20, 24–26, 29, 30, 33, 55, 60, 71, 151, 153, 167, 175, 181
    instability (MSI), 29–31
microscope(s), 132, 169
microscopic, 132, 137, 146
microscopy, 131, 144
mini-sequencing, 100, 164, 166, 168, 170–172
monoallelic, 27
mononucleotide, 30
monosomy, 180
mouse, 18
mouthwashes, 63
multiple displacement amplification (MDA), 4–7, 19–21, 33, 47, 99, 100–119, 122–128, 130, 132, 133, 141, 143–146, 150, 151, 153, 159, 165–168
multiplex PCR, 164, 165
multiplexing, 125

mutation, 1, 166
   analysis, 30, 34, 164
   assay(s), 124
   detection, 163, 167
   frequencies, 124
mutational analysis, 47

needle microdissection, 26
nonenzymatic fragmentation, 60
Nycodenz, 134, 136, 137

oligo(dT), 89
oligonucleotide
   adaptors, 2, 3
   arrays, 33
   microarrays, 145

paraffin-embedded, 24, 154
paralogs, 144
phenotype(s), 29, 144
phenotypic, 143, 144, 146
φ29, 4, 6, 33, 100–105, 108–110, 113, 115,
   119, 130, 150, 153
phosphorothioate, 4, 105
phylogenetic, 132, 137
PicoGreen, 14, 18, 19, 28, 63, 68, 82, 110,
   153
plasma, 107, 108
point mutations, 100
polar bodies, 163
poly(dT), 3, 80, 81, 84, 87, 96
polyA, 92, 97
polyacrylamide, 30, 103–106
polymerase chain reaction (PCR), 1–3,
   11–20, 23–25, 27, 29, 30, 32–37,
   39, 41–43, 51–53, 55, 57, 60, 62,
   64–71, 77, 79, 82, 91, 92, 99, 102,
   111, 113–115, 122–125, 127, 132,
   137, 144, 150–152, 159, 164,
   166–168, 171, 179
polymorphic, 30, 164, 167, 182
pre-implantation, 163, 173, 175–177,
   179–181
   genetic diagnostics (PGD), 3, 100, 163,
      167, 168, 174, 175, 179
primer extension, 24, 102, 103, 171
primer extension pre-amplification (PEP)
   PCR, 2, 11, 23–25, 47, 100,
   164–166
prokaryotic, 48, 132, 133, 144

proliferation, 57
proofreading, 6, 24, 105
prostate cancer, 117, 149
protruding ends, 81
pyrosequencing, 55

quantitative PCR, 6, 113, 116, 127, 129,
   130, 155, 159

random cleavage/shear, 3
random fragmentation, 7, 48, 60, 72
random hexamer(s), 5, 11, 23, 89, 102, 105,
   106, 109, 150, 153, 167, 170
random PCR (R-PCR), 77, 79, 83, 92, 93,
   97
random primer(s), 2, 4, 6, 28, 35, 100, 106,
   110, 113, 114
randomly fragmented, 79
real-time PCR (RT-PCR), 68, 114, 151, 153,
   154
recessed ends, 81
repetitive DNA, 6
replication
   bubble, 100
   fidelity, 108
   fork(s), 101, 102, 104
REPLI-g, 21, 109, 123, 141, 153, 168, 170
representational bias, 3
restriction and circularization aided RCA
   (RCA-RCA), 6, 7, 150, 151,
   153–159
restriction digest, 79, 96
restriction endonuclease(s)/enzyme(s), 3, 6,
   34, 35, 96, 150, 153, 155, 159
restriction fragment length polymorphism
   (RFLP), 100
reverse transcription, 79, 80, 89, 97, 158
ribosomal RNA (rRNA), 133, 137, 144
rolling-circle, 104
rolling-circle amplification (RCA), 4, 6,
   150, 151

second strand synthesis, 79, 80, 84, 87, 92,
   93, 96
sequence
   analysis, 16, 17, 25, 30, 32, 55, 69
   bias, 53
   data, 70
sequenced, 14, 30, 69, 70, 123, 127, 142

sequence-specific primer PCR (SSP–PCR), 12, 14, 17, 21
sequencing, 16, 21, 24, 31, 35, 60, 99, 100, 110, 111, 119, 120, 123, 125, 127, 132, 137, 138, 141–146, 164, 166, 171, 172,
serum, 107, 108
sheared DNA, 50, 51
shotgun sequencing, 143
single bacterial cells, 3, 7, 100
single cell comparative genomic hybridization (SCOMP), 3, 7, 33–36, 39–44
single cell(s), 2, 7, 23, 34, 35, 37, 39, 60, 121, 122, 124–132, 143–146, 151, 163, 164, 167, 169, 170, 173
single molecule fluorescence detection (SMFD), 14, 17
single nucleotide polymorphism (SNP), 1, 7, 11, 12, 14, 16–21, 55, 72, 99, 100, 111–113, 116, 124, 151, 167, 168
    arrays, 33
Southern blotting, 48, 56, 57
spectrophotometer, 15–17, 20, 27, 28, 38, 42, 63, 68, 81, 82
SpeedVac, 27, 28
strand displacement, 2, 4–7, 81, 87, 101–105, 130, 167
    amplification, 60
stromal, 26
    cells, 149

T4 gene 32 protein, 4
T7
    promoter, 79, 80
    RNA polymerase, 3
tandem repeat, 167
TaqMan, 111, 112, 114–116, 124, 126, 128–131, 153, 155, 158, 159
telomeres, 111, 176, 179, 180, 182
terminal deoxynucleotide transferase (TdT), 3, 79–81, 86, 91, 96
terminal transferase, 80, 84
tetranucleotide, 71, 176
thermal cycler/thermocycler, 15–17, 28, 37–41, 64, 65, 87–89, 109, 110, 122, 123, 170
thermocycling, 2
thermophiles, 142
threshold cycle, 114
tissue fixation, 75
tissue microdissection, 26
tissue section, 108
translocation, 167
trinucleotide, 71, 176
trisomy, 180
tumor(s), 1, 24–26, 30, 31, 72, 117, 149, 150
    cell(s), 25, 26, 34
    dissemination, 24
tumor protein p53 gene (TP53), 30

unculturable species, 100